Eleventh Edition

MATHEMATICS
ACTIVITIES

For Teaching
&
Learning

Jane Thompson Barnard
Ed R. Wheeler
Armstrong Atlantic State University

KENDALL/HUNT PUBLISHING COMPANY
4050 Westmark Drive   Dubuque, Iowa 52002

Book Team

Chairman and Chief Executive Officer   Mark C. Falb
Vice President, Director of National Book Program   Alfred C. Grisanti
Editorial Development Supervisor   Georgia Botsford
Developmental Editor   Angela Willenbring
Prepress Project Coordinator   Sheri Hosek
Prepress Editor   Michele Steger
Design Manager   Jodi Splinter
Designer   Suzanne Millius

TABLE OF CONTENTS

PREFACE

About this Book:

Mathematics Activities for Teaching and Learning provides a hands-on, problem-solving approach to learning and teaching the mathematics that needs to be known and understood by teachers of elementary and middle school mathematics. This laboratory text has been used to serve four audiences. It has been used to support workshops and short courses for in-service elementary and middle school teachers. It has been used as a primary text in methods classes for in-service elementary and middle school teachers. It has been used by elementary and middle school teachers as a resource in their classrooms. Finally, it has been used to support a laboratory component for content courses in mathematics directed at liberal arts majors and students who intend to major in education. To this end, *Mathematics Activities for Teaching and Learning* has been organized to accompany the text *Modern Mathematics* by Ruric E. Wheeler and Ed R. Wheeler, and the activities are grouped to correspond to the chapters of that text.

Principles and Standards of NCTM:

Throughout this edition, careful attention has been given to the wisdom from the *Principles and Standards for School Mathematics* published by the National Council of Teachers of Mathematics in 2000. As appropriate, material from the narrative portion of this publication has been referenced in the *Teaching Notes* that follow each section, and a listing of the Principles and Standards can be found in their entirety beginning with Appendix Page 33 and following.

General Goal:

The 2001 publication of the Mathematics Association of America, *The Mathematical Education of Teachers,* asserts that –

- Courses on fundamental ideas of school mathematics should focus on a thorough development of basic mathematical ideas. All courses designed for prospective teachers should develop careful reasoning and mathematical 'common sense' in analyzing conceptual relationships.

- Along with building mathematical knowledge, mathematics courses for prospective teachers should develop the habits of mind of a mathematical thinker and demonstrate flexible, interactive styles of teaching.

The goal of *Mathematics Activities for Teaching and Learning* and of the textbook *Modern Mathematics* is to provide the resources for creating such college courses. Students who learn mathematics using these resources will be more confident of their knowledge and better prepared to share it with others.

Specific Goal:

In the vision statement of the 2000 publication of the National Council of Teachers of Mathematics, *Principles and Standards for School Mathematics,* we find this statement:

" . . . Our students deserve and need the best mathematics education possible, one that enables them to fulfill personal ambitions and career goals in an ever-changing world."

The specific goal of this activities manual is directed at educating elementary and middle school teachers who will design and deliver this quality mathematics education. This text will provide hands-on, manipulative-based activities that will involve teachers in discovering and exploring concepts in interesting and real-world settings. By developing ideas from the concrete to the pictorial to the abstract, by learning concepts in the context of problem solving, and by considering multiple representations of mathematical ideas, prospective elementary and middle school teachers will be better prepared for the job ahead of them.

Acknowledgments:

We would like to express our appreciation to the following persons who helped us develop this book:

- Bill Kring and Aaron Wheeler who designed the format of this text, executed the art work, commented on the manuscript, and encouraged the authors.
- Ruric Wheeler who is responsible for eleven fine editions of *Modern Mathematics,* a text that has educated many tens of thousands of elementary school teachers.
- Dale Kilhefner who introduced us many years ago to activity-based teaching and who provided first drafts of several of these activities.
- Angela Willenbring of Kendall/Hunt Publishing Company who provided counsel in text preparation and guided the book through production.
- Susan Ouzts and Larry Lesser who teach regularly from this text and have suggested many improvements.
- Cindy Thompson Erwin who looked over the revised manuscript with a careful eye.
- The National Council of Teachers of Mathematics for permission to quote from the *Principles and Standards for School Mathematics* as published in 2000 and from the *Curriculum and Evaluation Standards* published in 1989.
- The Mathematics Association of America for permission to quote from *The Mathematical Education of Teachers.*

Communication:

We would appreciate any suggestions for improvement from those who use this text. Please write us at the Department of Mathematics, Armstrong Atlantic State University, 11935 Abercorn Street, Savannah, Georgia, 31419 or send us electronic mail at the following addresses: barnarja@mail.armstrong.edu or wheeleed@mail.armstrong.edu.

ABOUT THE AUTHORS

Jane Thompson Barnard, Associate Professor of Mathematics, received her B.S. and M.S. in Mathematics as well as Ed.S. (Middle Grades) from Georgia Southern University. After teaching secondary and middle grades mathematics, she came to Armstrong Atlantic State University where she has taught mathematics and mathematics education courses for twenty-two years. During that time, she has also taught elementary and middle grades students in special projects and continues to teach in K-12 schools on a regular basis. Jane was awarded the 2001 Regents' Teaching Excellence Award by the Board of Regents of the University System of Georgia and has received four teaching excellence awards from her university. She is active in the National Council of Teachers of Mathematics, having served as chair of *Student Math Notes* as well as in other leadership positions. She has also served as the president of the Georgia Council of Teachers of Mathematics and as Southern I Director for the National Council of Supervisors of Mathematics. For ten years, she was the lead teacher in MathStart, a summer program for middle schoolers. Jane has received over 25 Eisenhower Higher Education grants and conducts workshops and trainings all over the country – most recently in South Africa.

Dr. Ed R. Wheeler received his Ph.D. in mathematics from the University of Virginia in 1973. He taught at Northern Kentucky State University and Meredith College before coming to Armstrong Atlantic State University as head of the Department of Mathematics and Computer Science. Most recently, Dr. Wheeler assumed the position of Dean of the College of Arts and Sciences at AASU. Dr. Wheeler has been author or co-author of seven textbooks including serving as coauthor of the ninth, tenth, and eleventh editions of *Modern Mathematics* with R. E. Wheeler. For ten years, he was the Director of MathStart, a summer program for middle grades at-risk students. He was also awarded the Distinguished Teaching Award at Armstrong in 1998.

ORGANIZATION AND FEATURES

Chapter Introductions:

In short introductions to each chapter, the authors identify the significance of the material in that chapter to the elementary and middle school curricula and give an overview of the activities that will follow.

Activities:

The heart of each chapter is a set of activities. Most of the activities are manipulative-based and each relates to the content of the chapter of the textbook. Each of the activities begins with an *Overview* and a description of *Materials* needed. When appropriate, black-line masters for the materials are provided in the Appendix at the end of the manual; in most other cases the materials are items that are easily obtained. In particular, attractive sets of materials for many activities can be easily produced using an Ellison Letter Machine™ and in such cases the appropriate die will be identified. We will leave it to the instructor to determine the optimal grouping for many of the activities. In those circumstances in which a specific group size is mandated by the activity, this fact is indicated in the overview.

Teaching Notes:

Scattered among the activities are notes addressing pedagogical nuances of the material being discussed. These notes include material on how to help young learners master the concepts being discussed, common errors made in presentation of the concepts, and notes on how to modify activities to address the needs of different age groups.

Teaching Notes - A Summary:

At the end of each chapter, the authors establish the connection between the NCTM Principles and Standards and the material in the chapter. Further, they continue their conversation with prospective teachers about how to integrate the concepts and activities of the chapter into the classroom.

Resource Guide:

At the end of each chapter is a brief bibliography that includes recent publications from *Teaching Children Mathematics* (formerly *Arithmetic Teacher*), *Mathematics Teaching in the Middle School,* and *Mathematics Teacher*, other important books and articles related to the material being discussed, literature books which enhance the language arts connection to mathematics, and other resources such as video tapes and classroom materials. Addresses and phone numbers to contact in order to acquire non-print resources are found on the next page. A video resource list is found on Appendix Page 32.

RESOURCE ADDRESS LIST

Blue Marble Children's Bookstore
1356 S. Ft. Thomas Avenue
Fort Thomas, KY 41075
(859) 781-0602
FAX (859) 781-0728
www.bluemarblebooks.com

Pearson Learning/Dale Seymour
　Publications®
P.O. Box 2500
Lebanon, IN 46052
(800) 321-3106
FAX (800) 393-3156
www.pearsonlearning.com

Delta Education
P.O. Box 3000
Nashua, NH 03061
(800) 442-5444
FAX (800) 282-9560
www.delta-education.com

Ellison Educational Equipment, Inc.
(Ellison Letter Machine and dies)
25862 Commercentre Drive
Lake Forest, CA 92630-8804
(800) 253-2238
FAX (800) 253-2240
www.ellison.com

ETA/Cuisenaire®
500 Greenview Court
Vernon Hills, IL
(800) 445-5985
FAX (800) 382-9326
www.etacuisenaire.com

National Council of Teachers of Mathematics
1906 Association Drive
Reston, VA 20191-1502
(800) 235-7566
FAX (703) 476-2970
www.nctm.org

Key Curriculum Press
1150 65th Street
Emeryville, CA 94608
(800) 338-7638
FAX (510) 595-7040
www.keypress.com

Project MATHEMATICS! (videotapes)
California Institute of Technology Bookstore
Mail Code 1-51
Pasadena, CA 91125
(800) 514-BOOK
FAX (626) 795-3156
www.bookstore.caltech.edu\

The Wright Group/McGraw-Hill
19201 120th Avenue NE, Suite 100
Bothell, WA 98011
(800) 523-2371
FAX (800) 593-4418
www.thewrightgroup.com

1 CRITICAL THINKING AND PROBLEM SOLVING

To solve a problem is to find a way where no way is known offhand, to find a way out of difficulty, to find a way out around an obstacle, to attain a desired end that is not immediately attainable, by appropriate means. - *George Polya*

In the not so distant past, mathematics was treated in the elementary and middle school classrooms as a static collection of rules and facts used exclusively to complete an endless stream of arithmetic computations. Over the past decade and a half, a new dynamism has swept into those same mathematics classrooms. Though this dynamism has many sources, none is more important than the realization that problem solving should be the central focus of the teaching of mathematics.

> "Problem solving means engaging in a task for which the solution method is not known in advance. In order to find a solution, students must draw on their knowledge, and through this process, they will often develop new mathematical understandings. Solving problems is not only a goal of learning mathematics but also a major means of doing so. Students should have frequent opportunities to formulate, grapple with, and solve complex problems that require a significant amount of effort and should then be encouraged to reflect on their thinking. . . . Problem solving is an integral part of all mathematics learning, and so it should not be an isolated part of the mathematics program."
> -*Principles and Standards for School Mathematics*

The activities of this chapter are designed to help you develop your skills as a flexible thinker and problem solver. You will find activities built around many of the classic problem-solving paradigms: **look for a pattern; make a table; draw a picture; guess, test, and revise; form an algebraic model; try a simpler version; use reasoning;** and **use a variable.** These activities will provide you with a valuable foundation of experience as you continue to grow as a problem solver and as you move to helping children grow as problem solvers.

However, it is important to realize that the goal is not to mechanically apply a given "strategy" to a specific kind of problem. Rather, a good problem solver will be able to apply an array of problem-solving tools to solve "new" problems, problems that he or she has not yet encountered. For this reason you will find that Chapter 1 of the Wheelers' text and this chapter of the activities manual are just an introduction and invitation to problem solving. In the subsequent chapters of the text and this manual, you will be invited again to the problem solving table to improve your skills and prepare to lead your students to improve their skills.

Often in the activities you will be asked to verbally describe the pattern or strategy that you have discovered. Other times you will be asked to write your own problem and discuss its solution with a colleague. Mathematics is a language and, as such, needs to be practiced as written and oral communication. By making mathematics a part of the conversation of your class, both the one in which you presently learn and the one you will eventually teach, you will ensure that the students of both classes master the concepts as well as the computations that are being studied.

Activity 1-1: Sequences of Geometric Objects

Overview: An important critical thinking skill is the ability to identify and use patterns in a variety of different contexts. In this activity you will identify and describe patterns of sequences of geometric objects.

Discover a pattern in the way that the successive geometric figures are formed and fill the blanks for the next entries in the sequence. Then describe verbally the pattern that you are using in the space below the pattern.

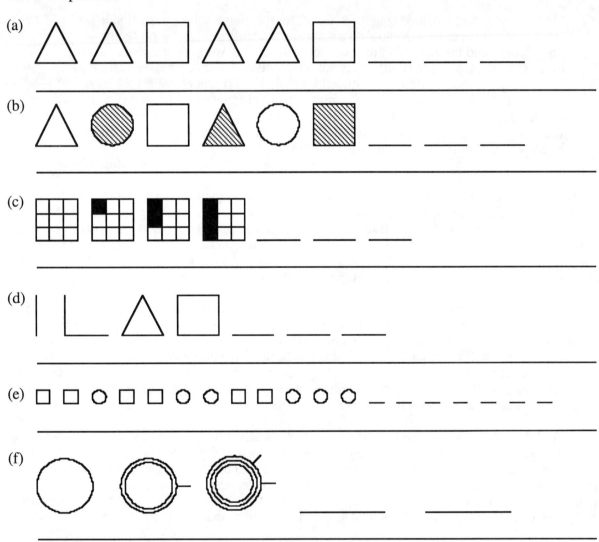

(g) Create two geometric sequences of your own and ask a colleague to discover the pattern.

(h) Make an alphabet sequence that used a pattern similar to the one found in (a); in (e).

Activity 1-2: Sequences of Numbers

Overview: An important critical thinking skill is the ability to identify and use patterns in a variety of different contexts. In this activity you will identify and describe patterns of sequences of numbers.

Discover a pattern in the way that the successive terms of the sequences are formed and find the next three entries in the sequence. In the space provided below each sequence describe verbally the pattern that you are using. Challenge Problem: Find the 50th term; the nth term.

(a) 1996, 1992, 1988, 1984, ____, ____, ____, · · ·

_____

(b) 73, 82, 91, 100, ____, ____, ____, · · ·

_____

(c) 1, 10, 100, ____, ____, ____, · · ·

_____

(d) 49, 45, 41, 37, ____, ____, ____, · · ·

_____

(e) 1, 4, 5, 1, 4, 5, 1, 4, 5, ____, ____, ____, · · ·

_____

(f) 1, 4, 9, 16, ____, ____, ____, · · ·

_____

(g) 3, 7, 12, 18, 25, ____, ____, ____, · · ·

_____

(h) Make up two number sequences and ask a classmate to discover the pattern, predict the next three terms, and describe the pattern.

Activity 1-3: Describing the Pattern

Overview: In this activity we will not only determine a pattern in the number sequences and use it to predict new elements, we will also describe the pattern.

A. Consider this sequence: 2, 5, 8, 11, 14, ____, ____, ____, . . .

(a) Fill in the blanks with the next three terms of this sequence.

(b) Describe in words how you formed each new term of this sequence from the previous term:
New term = Previous term + _____

Note: A sequence in which each term is formed from the previous term by adding a fixed number, the common difference, is called an *arithmetic sequence*. This common difference is also the *rate of constant increase* or the *rate of constant decrease*, depending on whether the terms of the sequence increase or decrease.

(c) Complete this table showing how the term number of this arithmetic sequence is related to the term itself.

Term Number	Term	Simplified Term
1	2 = 2	2 + 0(3)
2	5 = (2) + 3	2 + 1(3)
3	8 = (2 +3) + 3	2 + 2(3)
4	11 = (2 + 3 + 3) + 3	2 + 3(3)
5	14 =	2 + __(3)
6		

(d) In view of the table in part (c), what is a rule for forming the term of this arithmetic sequence corresponding to term number n (the nth term)?

(e) Write five terms of an arithmetic sequence and find a rule for the nth term.

B. Consider the sequence: 3, 6, 12, 24, ___, ___, ___, . . .

  (a) Fill in the blanks with the next three terms of this sequence.

  (b) Describe in words how you formed each new term of this sequence from the previous term.
      New term = Previous term × _____

      Note: A sequence in which each term is formed from the previous term by multiplying by a fixed nonzero number is called a *geometric sequence*.

  (c) Complete this table showing how the term number of this geometric sequence is related to the term itself.

Term Number	Term	Simplified Term
1	$3 = 3$	$3 \cdot 2^0$
2	$6 = (3) \cdot 2$	$3 \cdot 2^1$
3	$12 = (3 \cdot 2) \cdot 2$	$3 \cdot 2^2$
4	$24 = (3 \cdot 2 \cdot 2) \cdot 2$	
5	$48 =$	
6	$96 =$	

  (d) In view of the table in part (c), what is a rule for forming the term corresponding to term number n (the nth term)?

  (e) Make up your own geometric sequence. Write the first five terms and find a rule for the nth term.

Activity 1-4: Those Fascinating Numbers of Fibonacci

Overview: No number sequence has been studied more intensely than the number sequence first discussed by Leonardo Fibonacci, an Italian merchant/mathematician of the thirteenth century. His famous sequence

$$1, 1, 2, 3, 5, 8, 13, 21, ...$$

has been found to enumerate numerous patterns in nature (from whorls in pine cones and pineapples to chambers of the chambered nautilus) and to pop up in surprising places throughout mathematics. In this activity we will use inductive reasoning to unravel some of the patterns that are hidden in Fibonacci's sequence.

Materials: Calculator

A. We will use f_1, f_2, f_3, .. to represent the terms of the Fibonacci sequence. Determine the patterns used to form this sequence and complete the following table with the first 18 terms in the sequence. In the third row of the table compute the squares of the terms of the sequence.

(a)

Name of term	f_1	f_2	f_3	f_4	f_5	f_6	f_7	f_8	f_9
Term	1	1	2	3	5	8	13		
Square of term	1	1	4	9	25				

Name of term	f_{10}	f_{11}	f_{12}	f_{13}	f_{14}	f_{15}	f_{16}	f_{17}	f_{18}
Term									
Square of term									

(b) Describe in words how you formed the terms of the Fibonacci sequence.

(c) Describe, using the term names f_1, f_2, f_3, ..., how you formed the terms of the Fibonacci sequence. That is:

$$f_3 = f_1 + f_2$$
$$f_4 = f_2 + f_3$$
$$f_5 = f_3 + f_4$$

Describe the nth term,

$$f_n =$$

B. **Sums of Terms:** Consider the following sums from the Fibonacci sequence.

$1 + 1 = 3 - 1$
$1 + 1 + 2 = 5 - 1$
$1 + 1 + 2 + 3 = 8 - 1$
$1 + 1 + 2 + 3 + 5 = 13 - 1$

(a) Continue this investigation until you see the pattern.

(b) Use the pattern to compute the sum $1 + 1 + 2 + 3 + 5 + 8 + 13 + ... + 987$

(c) Use the pattern to compute the sum $f_1 + f_2 + f_3 + ... + f_n$

C. **Sums of the Squares of the Terms:** Consider the following sums of squares of terms.

$1^2 + 1^2 = 1 \times 2$
$1^2 + 1^2 + 2^2 = 2 \times 3$
$1^2 + 1^2 + 2^2 + 3^2 = 3 \times 5$
$1^2 + 1^2 + 2^2 + 3^2 + 5^2 = 5 \times 8$

(a) Continue this investigation until you see the pattern.

(b) Use the pattern to predict this sum and then check it with your calculator.
$$1^2 + 1^2 + 2^2 + 3^2 + 5^2 + ... + 377^2 =$$

(c) What is this sum? $1^2 + 1^2 + 2^2 + 3^2 + 5^2 + ... + f_{12}^2 =$

D. **Squares of Terms:** Look back at the third row of the table in Part A and consider the squares of the terms.

$f_1^2 + f_2^2 = 1 + 1 = $ _____
$f_2^2 + f_3^2 = 1 + 4 = $ _____
$f_3^2 + f_4^2 = $ _____ $= $ _____
$f_4^2 + f_5^2 = $ _____ $= $ _____

Now, identify each of the numbers in the right hand column as a Fibonacci number. Do you see a pattern? Which Fibonacci number should give the sum of the squares of f_5 and f_6? Check your conjecture.

Challenge Problem: What is this sum? $f_{n-1}^2 + f_n^2$

Activity 1-5: Draw a Picture

Overview: Drawing a picture, a graph, or a diagram will nearly always make the problem easier to understand. In some cases, the solution is immediate after the picture is drawn. Try this strategy on the following problem.

The Problem: Barely awake, Jason left his room and went in search of the breakfast bar of his hotel. He entered the elevator and went down 5 floors. When no breakfast was to be found, he went up 9 floors, then down 7 floors. On this floor he found the breakfast bar, but it was still closed. By now he had forgotten the floor on which his room was located. If the breakfast bar is on the third floor, can you help Jason find the floor where his room is located?

Activity 1-6: Guess, Test, and Revise

Overview: Making a guess and then working to verify (or disprove) that guess gives direction to your efforts at problem solving. Indeed some problems can be solved completely by guessing, testing your guess, and then making subsequent guesses on the basis of what you have learned. However, it is important to try to learn something about the correct answer from each incorrect guess that you test. Try the following problems using this strategy.

(a) William has 9 more Cincinnati Reds players represented in his baseball card collection than does Tony. Together they have 27 cards representing players from the Cincinnati Reds. How many cards for Reds players does each boy have? Record your guesses and revisions in the following table.

Will's Reds	Tony's Reds	Test of Guess
10	5	10 is not 9 more than 5; Is 10 - 5 = 9? Is 10 + 5 = 27?
12	3	12 is 9 more than 3. Is 12 + 3 = 27?

(b) Tony spent so much time taking care of his baseball card collection that he was careless with his arithmetic homework. To motivate his son to excellence, Tony's father offered to give him 8 cents for each correct answer, but added that he would deduct 4 cents for each incorrect answer. Alas, after two nights Tony had worked 54 problems but he and his father were even - neither owed the other any money. How many problems had Tony worked correctly?
Use the following table to record your guesses. Describe how you tested each guess.

Number Right	Number Wrong	Explain Your Test of This Guess

(c) As we use "Guess, Test, and Revise" in problems such as these, we often find ourselves establishing a pattern as we check our guesses. By recording that pattern using a variable, we can create an algebraic model for the problem. Look back at your guesses and tests in Part (a) and write the pattern that must be checked for a guess of R.

Will's Reds	Tony's Reds	Check
R	$R - 9$	

What is an algebraic model for this problem?

(d) Look carefully at the guesses and tests from Part (b). By recording the patterns you followed using a variable, find an algebraic model for this problem.

Number Right	Number Wrong	Check
X	$54 - X$	

Activity 1-7: Make a Chart or Table

Overview: An important way to gain insight about a problem involves organizing your investigation in a chart or table. You will find this strategy useful in the following problems.

The Problem: Two planes left Washington, D.C. for the 3,600-mile flight to London. Flight 1632 left at 6:00 p.m. and flew at a rate of 500 miles per hour. Flight 1111 left at 8:00 p.m. and flew at a rate of 900 miles per hour. About what time was it in Washington D.C. when Flight 1111 passed Flight 1632?

Complete the following table to help you estimate the time at which Flight 1111 overtook Flight 1632.

(a)

Time	Miles from Washington	
	Flight 1632	Flight 1111
6:00	0	0
7:00	500	0
8:00	1000	
9:00		
10:00		
11:00		
12:00		
1:00		

(b) From the chart we see that Flight 1111 passed Flight 1632 between _____ p.m. and _____ p.m. A reasonable estimate of when they passed would be ____:30 p.m.

(c) We used the relationship Distance = Rate × Time to solve this problem. In Part (b) we estimated the time at which the second airplane passed the first. Compute the distance that each plane had traveled at that estimated time.

Activity 1-8:   Try a Simpler Version:
Look for a Pattern Counting the Diagonals of a Polygon

Overview: We have seen that looking for a pattern is an underlying goal in all aspects of problem solving whether we are working with numerical sequences or problems posed verbally. A very productive strategy involves looking at simpler versions of complex problems and trying to determine a pattern from those simpler versions that will solve the original problem. Let us use this strategy to determine the number of diagonals of a polygon. Remember that a diagonal in a polygon is a segment connecting two non-adjacent vertices.

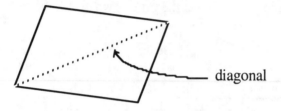
diagonal

The Problem: Determine the number of diagonals of a polygon with *n* edges.

If we wish to look at simpler cases, we might consider the triangle (three edges), the quadrilateral (four edges), and the pentagon (five edges). Count the number of diagonals in each case. Below each figure record the number of edges *n* and the number of diagonals *d*.

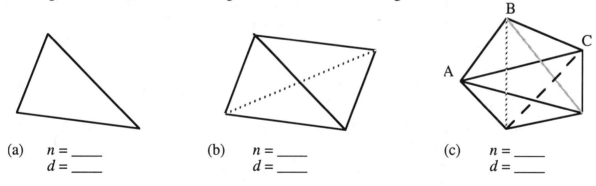

(a)   *n* = ____
       d = ____

(b)   *n* = ____
       d = ____

(c)   *n* = ____
       d = ____

Did you observe that the diagonals could be counted by counting the number of new diagonals that originated from successive vertices? For instance, in the pentagon there are 2 diagonals originating from vertex *A* (solid segments), 2 new diagonals from vertex *B* (dotted segment), and 1 new diagonal from vertex *C* (dashed segment). Hence, the number of diagonals in the pentagon is given by the sum $2 + 2 + 1$.

(d) In the hexagon (six-edged polygon) below, we see that there are 3 diagonals originating from vertex A. Sketch the rest of the diagonals of this hexagon and write the number of diagonals as a sum: 3 + _____ = 9.

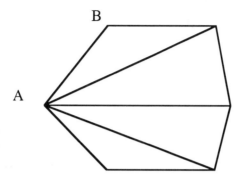

(e) Complete this table showing the sums used to compute the number of diagonals d for polygons with n edges.

$n = 3$	$d = 0$	$d = 0$
$n = 4$	$d = 1 + 1$	$d = 2$
$n = 5$	$d = 2 + 2 + 1$	$d = 5$
$n = 6$	$d = 3 + 3 + 2 + 1$	$d = 9$
$n = 7$	$d =$	$d = 14$
$n = 8$	$d =$	$d = 20$
$n = 9$	$d =$	$d = 27$

(f) Do you see the pattern? How many diagonals does a polygon with 15 edges have?

(g) Using this pattern, write an expression used to compute the number of diagonals of a polygon with n edges.

(h) **Challenge Problem:** Another expression used to compute the number of diagonals of a polygon with n edges is

$$\frac{n(n-3)}{2}.$$

Can you think of a way to view this problem to verify this formula?

Activity 1-9: Use Reasoning

Overview: Using reasoning in an organized and systematic way often yields solutions to problems. In the problems that follow you will find that using a chart to organize your information will allow you to eliminate impossible situations and recognize possible solutions.

A. Robert, Roland, Rudy, and Rufus are quadruplets. Grandpa Stuart can tell them apart only by the different color shirts that they wear. Roland and Rufus never wear green shirts, Rudy always wears a red shirt, and Roland started to choose a blue shirt but decided against it. Robert's favorite brother wears yellow. Record preferences on the following table and decide who is wearing what color shirt.

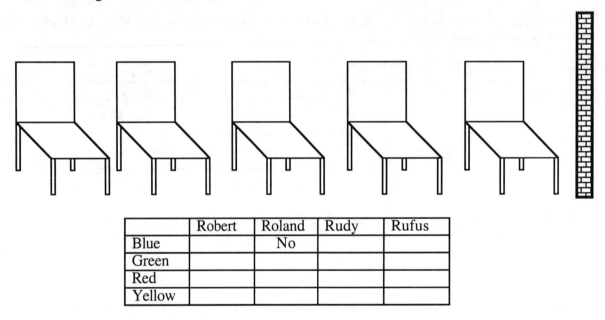

	Robert	Roland	Rudy	Rufus
Blue		No		
Green				
Red				
Yellow				

B. Five students sitting in a row are characterized by the given information. Use the information to determine their names, seating order, hometowns, and majors. As you read the information, you may find it useful to record all possible options on the backs of the chairs pictured below and then eliminate some of the options as more information is processed.

(a) Tim, the history major, sits next to the pre-med major.
(b) The student from Greenville sits in the middle between the music major and the student from Albany.
(c) Joy, in the fourth seat, bumps the art major's notebook when sitting down.
(d) Marie, against the wall, chats with the student from Miami.
(e) The student from Athens sits beside Carrie.
(f) Shane sits between the pre-med major and the Miami student.
(g) The end seats are occupied by a history major and a business major.
(h) Joy, a music major, sits next to an Oak Grove girl.

Activity 1-10: Work Backward

Overview: In some problems we have a clear view of what the result of a process is to be, and the problem is to understand the process as it develops. In this case, the advice of "work backward" is often good advice. Try these two problems using this advice.

A. In the days of the Great Depression, consumers were often quite ingenious in the deals they arranged with shopkeepers. A man entered a store and proposed to the shopkeeper, "If you give me as much money as I now have, I will spend $20 in your store." The owner agreed and the man spent the money. The man entered a second store and proposed the same deal. Again, the shopkeeper agreed and the money was spent. Yet again, the man entered a store, persuaded the shopkeeper, and $20 was spent. At this point the man had no money left. With how much money did he start?

   (a) When he left the third store, the man had no money. Let us determine how much he had when he entered the third store.

   How much did he spend in the third store? _____

   Of this, how much was money he brought with him?_____

   How much did he have when he entered the third store?_____

   (b) Use similar reasoning to determine how much he had when he entered the second store.

   With how much did he leave the second store?_____

   How much did he spend in the second store?_____

   How much did he have when he entered the second store?_____

   (c) With how much money did the man start?

B. On the fourth day of spring break, only the department head and Professors Munson, Hudson, and Matthews are laboring in their offices. Seeking to cheer the troops up, sweet Jane leaves a box of donut holes in the departmental office.
   At about the same time, miserable Mark drops by the departmental office hoping to wheedle a grade change from Professor Matthews. Spying the donut holes, Mark quickly stuffs six of them into his book bag. The department head comes in, divides the remaining "holes" into four equal piles, gives the two extras to Mark, and retreats to his office with his pile. Professor Munson comes by, carefully divides the remaining "holes" into four equal piles, gives Mark the extra donut hole, then returns to his office with his pile. Professors Hudson and Matthews come in together and divide the remaining donuts into four equal piles. They give the three extra donut holes to Mark for waiting patiently, they each take a pile of "holes" and Mark follows Professor Matthews to his office. As Mark leaves, he observes that there are a total of six donut holes remaining in the two piles.

   (a) How many donut holes were in each of the piles that were made by Professors Hudson and Matthews? _____

   (b) Now work backward to determine the number of "holes" that sweet Jane originally left in the office. _____

C. Consider the problem:

If a number is increased by 3 and that sum is multiplied by 2, the result is 20.

Whereas a beginning algebra student may find an answer by writing and solving the equation $2(N + 3) = 20$, an elementary student may use a schematic diagram. Nonetheless, the reasoning is basically the same and there is a nice correlation between the two.

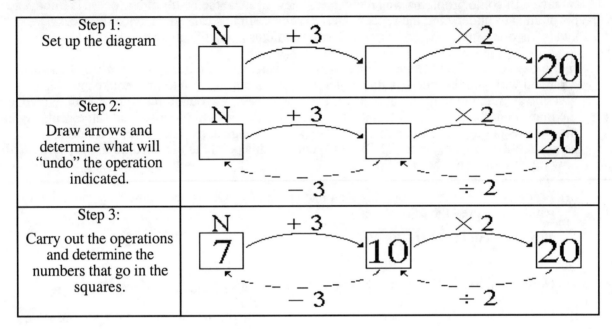

Algebraically, the process would look like this:

$$2(N+3) = 20$$
$$\frac{2(N + 3)}{2} = \frac{20}{2}$$
$$N + 3 = 10$$
$$N + 3 - 3 = 10 - 3$$
$$N = 7$$

There is a clear (some would say, isomorphic) relationship between the steps in the schematic diagram and the steps in solving the equation.

(a) Use a schematic diagram and the strategy of working backward to solve this problem:
   If a number is multiplied by 4 and 7 is added to that product, the result is 19.

(b) Use a schematic diagram and the strategy of working backward to solve this problem:
   If a number is increased by 5 and that sum is multiplied by $\frac{2}{3}$, the result is 8.

Activity 1-11: From Guess and Test to the Algebraic Model

Overview: Many algebra students both young and old are puzzled by how to form the algebraic equations that represent the problems that they study. They fail to grasp the fact that the algebraic equation merely reproduces the pattern of arithmetic used in a guess, test, and revise strategy of the problem. In the two activities that follow, we will examine problems proposed by middle grade friends of ours and use our work with guess and test to help us see how to use a variable to represent the problems.

A. **Jelana's Problem:** "In the driveway I see 73 wheels. Thirty children had arrived riding bicycles and tricycles. How many bicycles and how many tricycles are parked in the driveway?"

   (a) What do we know?

   How many wheels are on a bicycle? _____ How many wheels are on a tricycle? _____

   How many children? _____

   How many wheels are in the driveway? _____

   (b) Solve the problem using guess, test, and revise. Systematically record the arithmetic done in testing each guess.

Guess: Tricycles	Bicycles	Test of Guess
10	30 - 10	Is $3(10) + 2(30 - 10) = 73$?

   Write a sentence giving your solution to this problem.

B. Look carefully at the reasoning you used in testing your guesses. Use the pattern you developed in testing guesses and the variable T to find an algebraic model for this problem

Guess: Tricycles	Bicycles	Test of Guess
T	$30 - T$	

What equation represents the information in this problem?

C. **Jermonte's Problem:** "There were dogs and fleas. Together they had 50 heads and 240 legs. How many dogs and how many fleas?"

(a) What do we know?

How many heads does a dog have? _____ How many legs? _____

How many heads does a flea have? _____ How many legs? _____

(When Jermonte posed this problem, he had just learned in science that fleas are insects and, hence, have six legs.)

(b) Solve the problem using guess, test, and revise. Systematically record the arithmetic done in testing each guess.

Guess: Dogs	Fleas	Test of Guess

Write a sentence giving your solution to this problem.

D. Look carefully at the reasoning you used in testing your guesses. Use the pattern you developed in checking your guesses and the variable D to find an algebraic model for this problem.

Guess: Dogs	Fleas	Test of Guess
D		

What equation represents the information in this problem?

E. **Your Problem:**   Make up a problem for a colleague to solve. Solve the problem using guess, test and revise and create an algebraic model for the problem.

TEACHING NOTES: A SUMMARY

Teaching problem solving to young people will obviously take different forms depending on the age and maturity of the young people. The following statement from the *Standards (2000)* gives the flavor for problem solving as it will be taught to all students.

Instructional programs from prekindergarten through grade 12 should enable all students to—

- recognize reasoning and proof as fundamental aspects of mathematics;

- solve problems that arise in mathematics and in other contexts;

- apply and adapt a variety of appropriate strategies to solve problems;

- monitor and reflect on the process of mathematical problem solving.

"Posing problems comes naturally to young children: I wonder how long it would take to count to a million? How many soda cans would it take to fill the school building? Teachers and parents can foster this inclination by helping students make mathematical problems from their worlds. Teachers play an important role in the development of students' problem-solving dispositions by creating and maintaining classroom environments, from prekindergarten on, in which students are encouraged to explore, take risks, share failures and successes, and question one another. In such supportive environments, students develop confidence in their abilities and a willingness to engage in and explore problems, and they will be more likely to pose problems and to persist with challenging problems."

- Principles and Standards for School Mathematics

It is important in the early years of children's education that the problems they solve arise from school and other everyday situations. This will affirm to children that mathematics is connected in an essential way with their lives and experiences. Problems solved in early grades provide the foundation for concepts addressed in later grades. For example, understanding the pattern in simple cyclic sequences like 1, 4, 5, 1, 4, 5, ... (See Activity 1-2, (e)) in early grades prepares students to understand repeating, nonterminating decimal fractions like 0.145145... or $0.\overline{145}$ later.

In upper elementary and middle grades, students should solve problems from a much richer mathematics background. The foundation for probability, statistics, and geometry will have been laid in grades K - 4 and now this foundation should provide a rich resource for problems.

Do not forget that calculators and computers not only are good tools for use in problem solving, but also provide rich sources of problems to be solved. Your teaching of problem solving should provide opportunities for students to work cooperatively, use technology, and experience the power and relevance of mathematics.

RESOURCES

Periodicals:

Bledsoe, Gloria J. "Hook your Students on Problem Solving." *Arithmetic Teacher* 37 (December 1989): 16-19.

Boucher, Alfred C. "Critical Thinking through Estimation." *Teaching Children Mathematics* 4 (April 1998): 452-55.

Brown, Rebecca and Kristen Herbert. "Patterns as Tools for Algebraic Reasoning." *Teaching Children Mathematics* 3 (February 1997): 340-45.

Cobb, Paul, Erna Yackel, Terry Wood, Grayson Wheatley, and Graceann Merkel. "Creating a Problem-Solving Atmosphere." *Arithmetic Teacher* 36 (September 1988): 46–47.

Dossey, John A., Ina V. S. Mullis, and Chancey O. Jones. *Can Students Do Mathematical Problem Solving? Results from Constructed-Response Questions in NAEP's 1992 Mathematics Assessment.* 23-FR-01. Washington, D.C.: National Center for Education Statistics, August 1993.

English, Lyn D. "Promoting a Problem Solving Classroom." *Teaching Children Mathematics* 4 (November 1997): 172-79.

Ferrini-Mundy, Joan, Glenda Lappan, and Elizabeth Phillips. "Experiences with Patterning." *Teaching Children Mathematics* 3 (February 1997): 282-88.

Friedler, Louis M. "Problem Solving with Discrete Mathematics." *Teaching Children Mathematics* 2 (March 1996): 426-31.

Kroll, Diana Lambdin, and Tammy Miller. "Insights from Research on Mathematical Problem Solving in the Middle Grades." In *Research Ideas for the Classroom, Middle Grades Mathematics,* National Council of Teachers of Mathematics Research Interpretation Project, edited by Douglas T. Owens, 58–77. New York: Macmillan Publishing Co., 1993.

Loewen, A. "Creative Problem Solving." *Teaching Children Mathematics* 2 (October 1995): 96-99.

Nibbelink, William H. "Teaching Equations." *Arithmetic Teacher* 38 (November 1990): 48-51. *Middle-ground for mathematics instruction.*

Rosenbaum, Linda, Karla J. Behounek, Les Brown, and Janet V. Burcalow. "Step into Problem Solving with Cooperative Learning." *Arithmetic Teacher* 36 (March 1989): 7-11.

Sowder, Judith, and Larry Sowder. " Creating a Problem Solving Atmosphere." *Arithmetic Teacher* 36 (September 1988): 46-47.

Swenson, Esther J. "How Much Real Problem Solving?" *Arithmetic Teacher* 41(March, 1994): 400-03. *A contemporary view of problem solving.*

Verzoni, Kathryn A. "Turning Students into Problem Solvers." *Mathematics Teaching in the Middle School* 34 (October 1997): 102-7.

Williford, Harold. "Games for Developing Mathematical Strategy." *Mathematics Teacher* 85 (February 1992): 96-98.

Books & Other Literature:

Brown, Stephen I., and Marion Walter. *The Art of Problem Posing.* Hillsdale, N.J.: Lawrence Erlbaum Associates, 1983.

Burns, Marilyn. *A Collection of Math Lessons From Grades 1 Through 3* (1990). The Math Solution Publication (Distributed by Cuisenaire Company of America, Inc.).

Burns, Marilyn. *Fifty Problem-Solving Lessons, Grades 1–6.* Sausalito, Calif.: Math Solutions Publications, 1996.

Burns, Marilyn. *The Book of Think* (1976). Little, Brown and Company.

Burns, Marilyn, and Cathy McLaughlin. *A Collection of Math Lessons From Grades 6 Through 8* (1990). The Math Solution Publications (Distributed by Cuisenaire Company of America, Inc.).

Charles, Randall I., ed., and Silver, Edward A., ed. *The Teaching and Assessing of Mathematical Problem Solving* (1989). Lawrence Erlbaum Associates and NCTM.

Davidson, Patricia S., and Robert E.Willcutt. *Spatial Problem Solving with Cuisenaire Rods* (1983). Cuisenaire Company of America, Inc.

Polya, George. *How to Solve It* (1973). Princeton University Press.

Polya, George. "On Solving Mathematical Problems in High School," *Problem Solving in School Mathematics* (1980). NCTM.

Seymour, Dale, and Ed Beardslee. *Critical Thinking Activities in Patterns, Imagery, Logic.* [1988 (4-6), 1990 (K-3), 1991 (7-12)]. Dale Seymour Publications.

Other Resources:

Seymour, Dale. *Visual Thinking* (1983). Dale Seymour Publications. *Visualization Cards.*

2 — THE LANGUAGE OF LOGIC

Know not only the mathematics you teach, but also the mathematics that your students will learn. - anonymous

This is an important maxim for teachers from elementary school to college. To be an effective teacher you must understand where the mathematics you teach will reappear (in possibly more sophisticated form) in your students' education. Thus, it is important that a K - 4 teacher understand how the material he or she teaches will be used in grades 5-8, in high school, and in further education.

In studying Chapter 2 in the Wheelers' *Modern Mathematics* you are becoming proficient with some of the basic language and tools of logic and deductive reasoning. The ability to reason deductively becomes increasingly valuable as students progress through their education, and it is a crucial skill for life-long learning. Though the specific language of logic is not referenced until middle school, the foundation for deductive thinking reaches deep into the student's early education. As children in the early grades complete problem-solving exercises and puzzles requiring precise and systematic thinking, they are preparing to understand the language of logic, and more importantly, to become careful deductive thinkers. The activities in this chapter reflect the variety of ways that we prepare students to learn to reason well. In addition to the activities of this chapter, you will also want to give attention to the activities in Chapter 3 involving attribute pieces. These activities, accessible to children in many grades, also help prepare the foundations of logical thinking.

Activity 2-1: The Tower of Brahma

Overview: The Hindu legend tells it this way:
In the great temple at Benares, beneath the dome that marks the center of the world, God placed a brass plate in which are fixed three vertical diamond needles. When the world began, God placed 64 golden disks on one of the needles with the largest one laying on the brass plate and the others getting successively smaller up to the top one. This is the tower of Brahma. Each day the priests of the temple labor moving the disks according to the following rules:
   (i) Only one disk may be moved at a time.
  (ii) The disk that is moved must be placed on a needle so that there is no smaller disk below it.
When the 64 disks have all been moved from the original needle to another needle, the earth shall end.

Though we shall hope to avoid the dire consequences, we would like to determine how many moves are required to complete the process. We will simulate the process using the board on the following page and will use the strategies of solving a simpler problem and looking for a pattern.

Materials: The board found below and five coins of differing sizes (half dollar, quarter, nickel, dime, and penny) or five squares or circles of construction paper of descending sizes.

(a) Observe that three moves are required to solve the problem if the tower has two disks.

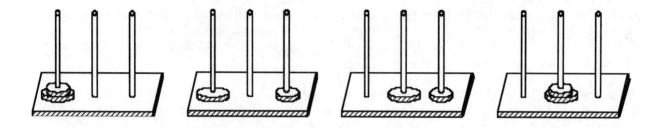

Use the board below and three of your disks to determine the number of moves required to move a stack of three disks from one place to another.

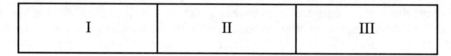

I	II	III

(b) Use four of your disks and determine how many moves are required to solve the problem if the original stack contains four disks.

(c) How many moves are required to solve the problem when the stack has five disks?

(d) Complete the following table:

Number of Disks	Number of Moves
1	$1 = 2^1 - 1$
2	$3 = 2^2 - 1$
3	$7 = 2^3 - 1$
4	$15 = 2^4 - 1$
5	
6	

(e) If we have 10 disks and each move takes 1 second, how long will it take to solve the puzzle?

(f) If the priests of the temple at Benares take 1 second to move a disk, how long will it take them to move all 64 disks?

Activity 2-2: Toothpick Teasers

Overview: In solving these teasers remember to be a flexible thinker: "Don't impose unnecessary conditions on the solution."

Materials: A box of toothpicks. (Note: This exercise might be done without toothpicks, but working with manipulatives encourages a spirit of experimentation that is often absent when thinking is confined to pencil and paper.)

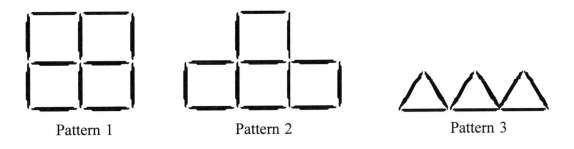

Pattern 1   Pattern 2   Pattern 3

(a) Remove 2 toothpicks from Pattern 1 leaving 2 squares.

(b) Remove 1 toothpick from Pattern 2 leaving 3 squares.

(c) Reposition 2 toothpicks in Pattern 1 to make 7 squares.

(d) Reposition 3 toothpicks in Pattern 3 to make 5 triangles.

Activity 2-3: Preparing for Deduction

Overview: In completing the following exercises borrowed from *Exploring Mathematics, Grade 8* (published by ScottForesman in 1991), you will strengthen your deductive skills and get a clearer view of the direction that must be given to the preparation of younger children.

A. At a costume party Julio, Bert, Neil, and Larry asked Lin to guess which one of them was dressed as an astronaut, a surgeon, a clown, and a monster. He recognized Julio's voice as the monster's. Bert was not the surgeon and the astronaut was neither Larry nor Bert. Summarize the facts in a table. Place a + in the table when you know who someone is. Put an *X* in the table when you know someone isn't in that costume.

	Astronaut	Surgeon	Clown	Monster
Julio				
Bert				
Neil				
Larry				

B. Perkins, Smythe, Dinwiddie, Bloom, and Magoo are an airline pilot, a sea captain, a bus driver, a rodeo rider, and an engineer, though not necessarily in that order. In their spare time one reads mystery novels, one studies nuclear physics, one writes poetry, one composes symphonies, and one plays the ukulele.

   Match each person with his profession and his pastime. Use the logic boxes on the following page to help you unravel the following clues. If a clue allows you to eliminate a possible combination, write *No* in the correct box or boxes. Write *Yes* if a clue confirms a combination. A *Yes* entry means that all other boxes in that row and that column can be filled in with *No*.

(a) The pilot is either Smythe or Bloom.
(b) Dinwiddie knows nothing about music or poetry.
(c) Magoo once played Parcheesi with the sea captain, the rodeo rider, and the engineer.
(d) The first poem the bus driver wrote was "My Parents."
(e) Neither Bloom nor Perkins writes symphonies, and neither Perkins nor Dinwiddie reads novels.
(f) The rodeo rider plays his ukulele while busting broncos.
(g) The man who studies nuclear physics gets seasick.
(h) Smythe is the pilot's brother-in-law.

	Pilot	Sea Captain	Bus Driver	Rodeo Rider	Engineer
Perkins					
Smythe					
Dinwiddie					
Bloom					
Magoo					

Activity 2-4: The Land of Truthtellers and Liars

Overview: By careful reasoning and giving attention to all possible cases, we will see if we can help poor Jack with his dilemma.

When Jack got to the top of the beanstalk, he found that there were two clans of giants. The Truthtellers were giants who always told the truth; the Liars were giants who never told the truth.

(a) Shortly after arrival, Jack met two giants. The first claimed that they belonged to different clans; the second declared that the first was a Liar. Help Jack determine to which clan each giant belonged.

(b) On another occasion, Jack fell into conversation with three giants. The first declared, "All three of us are Liars." The second objected, "Only two of us are Liars." The third then objected, "No, only one of us is a Liar." Again, to which clan did each belong?

(c) Jack's closest friends in the land beyond the beanstalk were Tragan, Sagan, and Fagan, the three children of the union of a Truthteller and a Liar. Individually, the children did not exhibit the characteristics of either Truthtellers or Liars. However, they had an annoying habit - when a question was asked of them, two would tell the truth and one would lie. Jack became very good at deciphering their conversation. Can you determine who was born first from the following responses?

Tragan: Sagan was born first.
Sagan: I am not the oldest.
Fagan: Tragan is the oldest.

TEACHING NOTES: A SUMMARY

Although students are usually in middle school before they learn the language of logic, reasoning is emphasized throughout the elementary curriculum. Indeed, "Mathematics as Reasoning" is the third process standard of the *Principles and Standards for School Mathematics*. However, in teaching reasoning to younger children, we must keep in mind that they are not ready for a rigorous development of logic rules. Rather, we must encourage them to explain their reasoning processes in their own words and ask them to support their ideas with pictures, diagrams, manipulatives, or technology whenever possible. From *Standards (2000)*:

Instructional programs from prekindergarten through grade 12 should enable all students to—

- recognize reasoning and proof as fundamental aspects of mathematics;

- make and investigate mathematical conjectures;

- develop and evaluate mathematical arguments and proofs;

- select and use various types of reasoning and methods of proof.

"From children's earliest experiences with mathematics, it is important to help them understand that assertions should always have reasons. Questions such as "Why do you think it is true?" and "Does anyone think the answer is different, and why do you think so?" help students see that statements need to be supported or refuted by evidence. Young children may wish to appeal to others as sources for their reasons ("My sister told me so") or even to vote to determine the best explanation, but students need to learn and agree on what is acceptable as an adequate argument in the *mathematics* classroom. These are the first steps toward realizing that mathematical reasoning is based on specific assumptions and rules.

Young children will express their conjectures and describe their thinking in their own words and often explore them using concrete materials and examples. Students at all grade levels should learn to investigate their conjectures using concrete materials, calculators and other tools, and increasingly through the grades, mathematical representations and symbols. They also need to learn to work with other students to formulate and explore their conjectures and to listen to and understand conjectures and explanations offered by classmates."
- *Principles and Standards for School Mathematics*

Sorting activities, ranging from the very simple to the very complex, can be used with students of all ages to develop reasoning skills. Examples of such activities will be found in Chapter 3.

Learning the language of logic is but one of the many ways students grow in their abilities to communicate mathematically. The Communication process standard indicates that reflection and communication are intertwined processes in mathematics learning. We often do not recognize oral and written communication about mathematical ideas as being an important and critical part of mathematics education . . . after all, mathematics is most often conveyed in symbols. From the *Principles and Standards for School Mathematics:*

Instructional programs from prekindergarten through grade 12 should enable all students to—

- organize and consolidate their mathematical thinking through communication;

- communicate their mathematical thinking coherently and clearly to peers, teachers, and others;

- analyze and evaluate the mathematical thinking and strategies of others;

- use the language of mathematics to express mathematical ideas precisely.

"Communication is an essential part of mathematics and mathematics education. It is a way of sharing ideas and clarifying understanding. Through communication, ideas become objects of reflection, refinement, discussion, and amendment. The communication process also helps build meaning and permanence for ideas and makes them public. When students are challenged to think and reason about mathematics and to communicate the results of their thinking to others orally or in writing, they learn to be clear and convincing. Listening to others' explanations gives students opportunities to develop their own understandings. Conversations in which mathematical ideas are explored from multiple perspectives help the participants sharpen their thinking and make connections. Students who are involved in discussions in which they justify solutions—especially in the face of disagreement—will gain better mathematical understanding as they work to convince their peers about differing points of view (Hatano and Inagaki 1991). Such activity also helps students develop a language for expressing mathematical ideas and an appreciation of the need for precision in that language. Students who have opportunities, encouragement, and support for speaking, writing, reading, and listening in mathematics classes reap dual benefits: they communicate to learn mathematics, and they learn to communicate mathematically."
- *Principles and Standards for School Mathematics*

The activities of this chapter can be used to enhance the reasoning skills of middle school students and, with modifications, can be used with upper elementary school children. The formal language of logic begins to appear in the middle school curriculum as an enrichment topic.

RESOURCES

Periodicals:

Hanna, Gila, and Erna Yackel. "Reasoning and Proof." In *A Research Companion to NCTM's Standards,* edited by Jeremy Kilpatrick, W. Gary Martin, and Deborah Schifter. Reston, Va.: National Council of Teachers of Mathematics, 2002.

Lampert, Magdalene, and Paul Cobb. "Communications and Language." In *A Research Companion to NCTM's Standards,* edited by Jeremy Kilpatrick, W. Gary Martin, and Deborah Schifter. Reston, Va.: National Council of Teachers of Mathematics, 2002.

Monchamp, Susan S. "Using Logic to Produce a Class Constitution." *Arithmetic Teacher* 40 (November 1992): 174-176.

Olson, Judith and Melfried Olson. "Classification and Logical Reasoning." *Teaching Children Mathematics* 4 (September 1997): 28-29.

Rubenstein, Rheta N., and Denisse R. Thompson. "Learning Mathematics Vocabulary: Potential Pitfalls and Instructional Strategies." *Mathematics Teacher* 93 (October 2000): 568-574.

Rubenstein, Rheta N., and Randy K. Schwartz. "Word Histories: Melding Mathematics and Meanings." *Mathematics Teaching in the Middle School* 6 (November 2000): 664-669.

Rowan, Thomas E., ed., Joe Garofalo, prep., and David K. Mtetwa, prep. "Mathematics as Reasoning." *Arithmetic Teacher* 37 (January 1990): 16-18.

Schwartz, James E. "'Silent Teacher' and Mathematics as Reasoning." *Arithmetic Teacher* 40 (October 1992): 122-124.

Witherspoon, Mary Lou. "And the Answer Is...Symbolic Literacy." *Teaching Children Mathematics* 5 (March 1999): 396-399.

Books & Other Literature:

Balka, Don. *Attribute Logic Block Activities* (1985). Ideal School Supply Company.

Botterill, D., and T. Durnin. *An Introduction to Logic Working with Attribute Materials* (1978). Invicta Educational Division.

Seymour, Dale, and Ed Beardslee. *Critical Thinking Activities in Patterns, Imagery, Logic* [1988 (4-6), 1990 (K-3), 1991 (7-12)]. Dale Seymour Publications.

Sherard III, Wade H. *Logic Number Problems* (1987). Dale Seymour Publications.

Thornburg, Pamela, and David Thornburg. *The Thinker's Toolbox* (1989). Dale Seymour Publications.

3   SETS, RELATIONS, AND FUNCTIONS

The view of mathematics as a dynamic subject must be begun as soon as children enter school. It does not take long for students to adopt the opposite view as they become receivers of procedures: "Tell me what to do, not why." The short-term payoff for students' knowing what to do is great, because that is what we reward. The long-term payoff is disaster, as shown by the present state of mathematical learning. - Mary M. Lindquist

Many of the important topics of elementary mathematics involve the study of sets. Indeed, the modern presentation of geometry and probability depends heavily on the language and concepts of sets. Prior to being able to use sets effectively students must learn to recognize the properties that determine whether or not an element is in a set. Sorting and classifying are two of the skills that prepare students to use the language of sets.

Attribute blocks are excellent tools for developing skills in sorting and classifying and for understanding the fundamental concepts of sets. A number of the activities of this section are built around attribute blocks. Although there are several slightly different sets of attribute blocks available, in our activities we will reference a set consisting of 32 blocks. In each set there are blocks of two sizes (large and small), four colors (blue, green, yellow, and red), and four shapes (triangle, rhombus, square, and circle). Each attribute block is named using three names: size - color - shape. For example, "small, green, rhombus" would obviously identify exactly one block in our set. We will represent the blocks in the activities as indicated in the examples below.

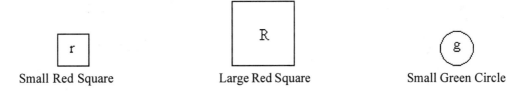

| Small Red Square | Large Red Square | Small Green Circle |

In activities involving sorting and classifying, talk with others in your group about how you are making your decisions. In communicating your reasoning, you will not only be improving your own ability to reason, you will be inviting others to share your thinking. There are three positive results of this invitation. The others in your group will possibly learn from you. In response, your classmates may share their thinking and you will learn from them. Together you can help one another avoid errors in reasoning. Learning to issue an invitation to others to share your reasoning is a mark of an excellent teacher.

TEACHING NOTES:
1. Conversation about attribute pieces provides opportunity for clarifying important geometric concepts. "Every square is a rhombus, but *square* is a more restrictive name than *rhombus* because each angle of a square is a right angle. Thus, a square is a special kind of rhombus."
2. If commercial manipulatives are not available for each student in a class, sets of attribute pieces can be cut using the Ellison Letter Machine™ and the "Attribute Pieces" die.

Activity 3-1: Sorting and Classifying

Overview: Skills at sorting and classifying are fundamental to mathematical thinking as well as thinking in the natural and social sciences. In this activity we practice the most fundamental of the sorting and classifying skills.
Students should work in pairs to complete these activities.

Materials: A complete set of attribute blocks and four twenty-four-inch pieces of yarn.

(a) Use two pieces of yarn to make two loops as shown below. Place a small red triangle in one loop and a large red triangle in the other. Select an attribute block from the set and place it in the loop "where it belongs." Distribute the rest of the attribute blocks into the two loops in such a way that all blocks in each loop share a common attribute.

In this exercise we classified the attribute blocks according to what attribute? _____

(b) Form four loops and place a large yellow triangle, a large yellow square, a large yellow circle, and a large yellow rhombus, respectively, in the four loops. Distribute the set of attribute blocks as in Part (a).

      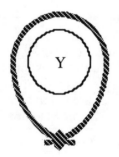   

In this exercise we classified the attribute blocks according to what attribute? _____

(c) Form four loops and place a small green triangle, a small blue triangle, a small yellow triangle, and a small red triangle, respectively, in the four loops. Distribute the set of attribute blocks as in Part (a).

In this exercise we classified the attribute blocks according to what attribute? _____

(d) Important note: The relation "has the same color as" is an equivalence relation on the set of attribute blocks. An equivalence relation on a set always partitions the set into disjoint nonoverlapping pieces. In Part (c) you demonstrated the partition of the set of attribute blocks produced by this relation. In Part (b) you created the partition of the set of attribute blocks produced by the relation _____.

TEACHING NOTE:

When beginning to use attribute blocks with children, particularly with younger children, place pairs of blocks on the overhead projector (or bulletin board) and ask, "How are they alike, how are they different?" For example, with these blocks,

Are they alike or different in:
- size? (Answer: Different, one is large, the other is small)
- color? (Answer: Alike, both are red)
- shape? (Answer: Alike, both are circles)

We say that these blocks are "one attribute different." With a similar analysis you can help children discover and articulate that the pieces below are "two attributes different."

Activity 3-2: Deliberating about Differences

Overview: Learning to identify characteristics that objects share and the characteristics that they do not share is a critical first step in understanding mathematical definitions. We will focus on this skill in this activity.

Materials: A set of attribute blocks

A. Place one attribute block in each circle of the diagrams below. The attribute blocks should be chosen so that the blocks in circles connected by arrows differ by exactly the number of attributes as there are arrows. For instance, in the first diagram we have placed a large red triangle and a small green triangle in adjacent circles connected by two arrows.

(a)

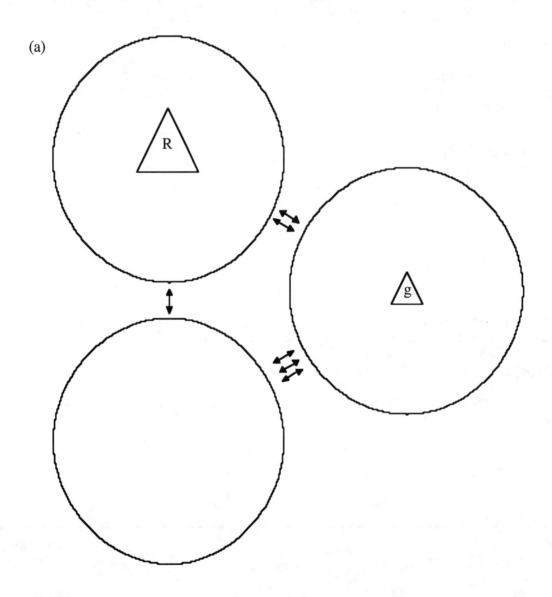

(b)

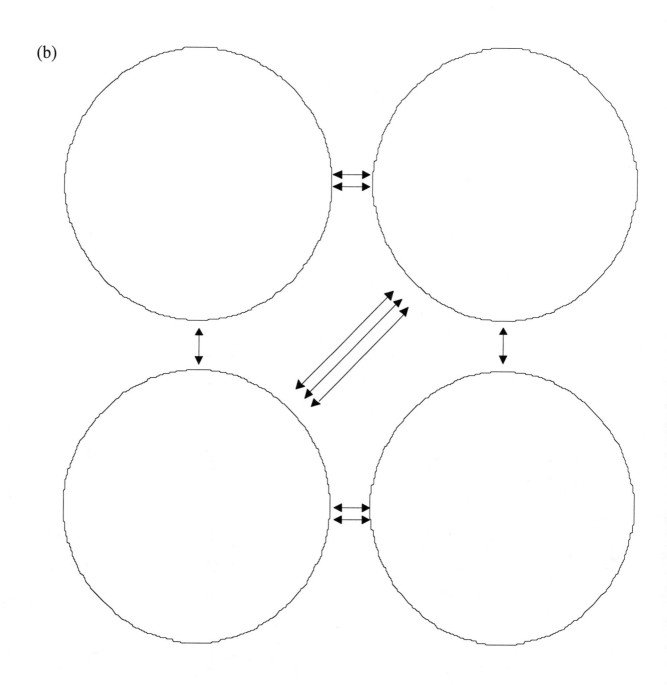

B. Place one attribute block in each of the squares of this diagram. Choose the blocks in such a way that blocks in adjacent squares differ by the number of attributes indicated on their adjacent edge.

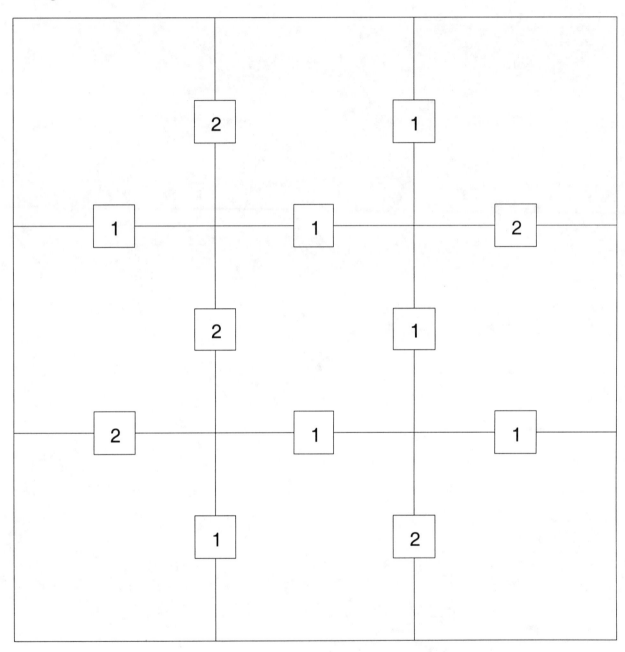

Activity 3-3: Venn Diagrams and Attributes

Overview: The notions of union and intersection of sets can be introduced and reinforced using attribute blocks and Venn diagrams.
This activity should be completed in pairs.

Materials: A set of attribute blocks and the large Venn diagrams found on Appendix Pages 1 and 2.

A. Copy the Venn diagram on Appendix Page 1. Label the Venn diagram as seen below and place each of the attribute blocks where it belongs in the diagram.

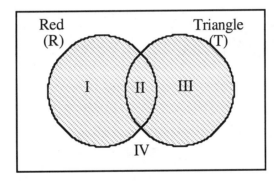

(a) Describe the attributes of the blocks in the shaded region in terms of the words "red" and "triangle."

(b) Describe the shaded region in terms of sets *R* and *T*.

(c) Describe the attributes of the blocks that lie in each of the regions I, II, III, and IV in terms of the words "red" and "triangle."

I:

II:

III:

IV:

(d) Describe each of the regions I, II, III, and IV in terms of the sets R and T.

I:

II:

III:

IV:

B. Copy the large Venn diagram on Appendix Page 2. Label the Venn diagram as seen below, and place each attribute block where it belongs.

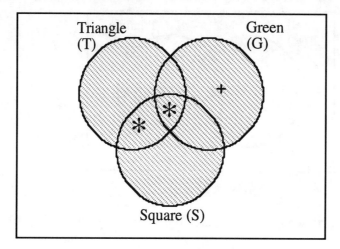

(a) How many blocks did you place in the regions marked with an asterisk?

What special name do we have for sets with this property?

Describe this set of blocks in terms of its attributes.

(b) Describe the attributes of the blocks in the shaded region.

(c) Describe the shaded region in terms of sets S, T, and G.

(d) Describe the attributes of the blocks in the region marked with a +.

Activity 3-4: Mysterious Missing Blocks

Overview: In this activity we reinforce the importance and significance of the words "and," "or," and "not" and build reasoning skills.

Materials: A set of attribute blocks [32 blocks: two sizes (large and small), four colors (blue, green, yellow, and red), and four shapes (triangle, rhombus, square, and circle)]. Recall that these can be cut using the Ellison Letter Machine™ and the "Attribute Pieces" die.

Discover the attribute block determined by each of the following sets of clues.

(a) It is green.
   It is not a triangle.
   It has 4 edges.
   It is not a square.
   It is small.

(b) It is large.
   It is not 4-edged.
   It is red.
   It is neither a triangle nor a rhombus.

(c) It is red.
   It is not large.
   It is not a circle.
   It has fewer than 4 edges.

(d) It is large.
   It is not green.
   It is not yellow and it is not red.
   It has 4 edges but it is not a square.

(e) It is neither a square nor a triangle.
   It is not round.
   It is small.
   It is green.

(f) It is small or it is yellow.
   It is not a circle.
   It is a triangle or it is a square.
   It is not large.
   It is red or green.
   It is yellow or it is red.
   It is not a triangle.

Activity 3-5: A Game: Label the Loop

Overview: In this game we will use indirect reasoning to guess the attributes of blocks that can be grouped together. Play the game with groups of three to five people.

Materials: A set of attribute blocks, a paper bag, and copies of Appendix Pages 1 and 2.

One player, the Game Master, will label each of the circles of the Venn diagram and cover the labels. The Game Master then hands the paper bag containing the attribute blocks to the other players. In turn, each of the other players blindly chooses an attribute block from the paper bag and tries to place it in one of the regions in the Venn diagram. The Game Master will then tell the player whether or not the attribute block can go in that region. If the block meets the requirement of the labels, it remains in the region; if not, it is removed and replaced in the bag. Then the next player chooses a block and tries to place it. The game ends when a player correctly guesses the labels on the circles.

(a) Use the Venn diagram from Appendix Page 1 and play the game with the labels seen below. The Game Master must be sure to cover the labels so that the other players cannot see them.

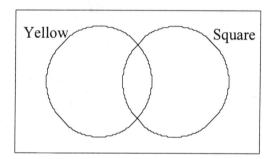

(b) Play the game three more times. For each new game, let a new Game Master determine the labels and cover them.

(c) Play the same game on the Venn diagram from Appendix Page 2. In this case all three circles must be labeled.

Activity 3-6: Guess My Rule

Overview: We will use our skills at conjecturing and recognizing patterns to learn the rules that govern the following functions. The column on the left holds the inputs; the column on the right holds the corresponding outputs.

(a)

x	$f(x)$
1	2
2	4
3	6
4	8
5	10

$f(x) =$ _____

(b)

x	$f(x)$
2	4
1	1
4	16
5	25
3	9

$f(x) =$ _____

(c)

x	$f(x)$
2	5
1	3
4	9
5	11
3	7

$f(x) =$ _____

(d)

x	$f(x)$
1	2
2	5
3	8
4	11
5	14

$f(x) =$ _____

(e)   Now make up two rules of your own and complete the corresponding tables. Ask a
        classmate to guess your rules.

x	$f(x)$

x	$f(x)$

TEACHING NOTE:

The Game of "Guess My Rule" can be used with children of many different ages with rules
that are chosen of appropriate complexity. Until later in middle school, however, expect the
children to communicate the rule verbally rather than using specialized notation like $f(x)$.
Even children as young as first or second grade can profit from activities such as this one if
they have toy cars to use if needed. Complete the activity below.

	1	2	3	4	5
	4	8			

TEACHING NOTES: A SUMMARY

As with logic, the formal language of sets is not learned until middle school. However, the groundwork preparing the student to use the conceptual framework of sets is laid throughout the elementary curriculum. In the early years the focus needs to be on sorting and classifying. Appropriately modified versions of the activities that we have completed using attribute blocks would be useful for this purpose. For younger children sorting activities like the one found in Parts (a), (b), and (c) of Activity 3-1 are appropriate. If a single loop is used, children can begin to play "Label the Loop" (Activity 3-5) as early as Grade 1 and complexity can be added for later grades by increasing the number of loops.

Older children should begin to attend to more complex sorting and classifying tasks. The tasks in Activity 3-2, scaled to an appropriate level of difficulty, would be useful in helping them focus on properties shared by all elements of a set, no elements of a set, and many elements of a set. One variation on these types of activities would include making a "train" in which the first person chooses an attribute block and each successive person getting in line would choose an attribute block that differs by one attribute from the block held by the preceding person. Certainly Activity 3-4 on "Mysterious Missing Blocks" would help expand both the reasoning and classifying skills of older elementary children. Activity 3-3 will help middle school children learn the rudiments of set theory and begin to use Venn diagrams. "Guess my Rule" (Activity 3-6) both fosters the spirit of conjecture and prepares students for the concept of function, a concept that undergirds most successful applications of mathematics.

> ABOUT CHANGE: In times of change learners inherit the earth, while the learned find themselves beautifully equipped to deal with a world that no longer exists.
> -   Eric Hoffer

RESOURCES

Periodicals:

Geer, Charles P. "Exploring Patterns, Relations, and Functions." *Arithmetic Teacher* 39 (May 1992): 19-21. *"Mathemagical" activities.*

Holly, Karen A. "Patterns and Functions." *Teaching Children Mathematics* 3 (February 1997): 312-13.

Quinn, Robert J. "Developing Conceptual Understanding of Relations and Functions with Attribute Blocks." *Mathematics Teaching in the Middle School* 3 (Nov.-Dec.1997): 186-90.

Willoughby, Stephen S. "Functions from Kindergarten through Sixth Grade." *Teaching Children Mathematics* 3 (February 1997): 314-18.

Books and Other Literature:

Seymour, Dale, and Ed Beardslee. *Critical Thinking Activities in Patterns, Imagery, Logic* [1988 (4-6), 1990 (K-3), 1991 (7-12)]. Dale Seymour Publications.

4

WHOLE NUMBERS AND NUMERATION

"Without an understanding of number systems and number theory, mathematics is a mysterious collection of facts. With such an understanding, mathematics is seen as a beautiful cohesive whole."
- NCTM *Curriculum and Evaluation Standards, 1989*

The understanding of number meanings comes gradually to children. They must first manipulate concrete objects and explain their thinking using their own oral language. Through this, children construct their own number meanings as they reflect on their actions. Written symbols must be linked to physical and pictorial models and oral names as students develop place value concepts.

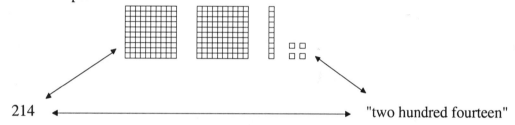

Counting on, counting back, and skip counting indicate understanding of number. Place value meanings grow out of grouping. Even though place value involves groups of ten, it is important for young children to look at grouping in general. If a student counts out 17 colored chips, he should be able to determine "How many groups of 5? How many left over?" or "How many groups of 10? How many left over?"

It is imperative for students to have opportunities to develop operation sense. As described in the *Standards*, operation sense means recognizing conditions in real-world situations that indicate the operation would be useful in those situations, building an awareness of models and properties of an operation, seeking relationships among operations, and acquiring insight into effect of an operation on a pair of numbers. Understanding the fundamental operations of addition, subtraction, multiplication, and division is central to knowing mathematics (*Standards*, 1989). Before computation, however, students must be introduced to the meaning of operations, concepts, and relationships. Informal experiences with all four operations must be begun in kindergarten and be continually provided through grade four. Some examples follow.

- Cassie, Lyn, and Huy each have two teddy bears. How many do they have when they place all of them together?

- Mario has 6 blocks and Shawn has 4 blocks. How many more blocks does Mario have than Shawn? or How many fewer blocks does Shawn have than Mario?

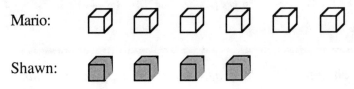

- Tricia, Bob, and Dick want to share 12 chocolate chip cookies equally. How many cookies does each one get?

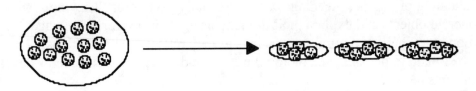

The language of basic operations: sum, addend, difference, product, factor, multiple, quotient, and divisor should be introduced orally and used informally as concepts of operations are developed. The activities in this chapter are representative of the many that will allow students to construct their own knowledge base.

Activity 4-1: Representing Numbers with Base Ten Pieces

Overview: Base ten blocks provide opportunities to experience and utilize place value concepts through physical manipulation. Each piece — the unit or one, the tens rod, the hundreds flat, the thousands cube — communicates through visual and tactile senses the important meaning of each place value position. When the base ten pieces are metric there is added value in that we have a model for metric length, area, and volume.

Work in groups of three.

Materials: Base ten pieces (commercial or those cut with the Ellison Letter Machine™) and three base ten dies

A. Each individual should find and record all possible ways to represent the numbers 46 and 105 using base ten pieces (hundreds, tens, and units).

TEACHING NOTE:
Often children will represent a number using different base ten pieces. For example, one child may represent 32 using 32 units, another may use 3 longs and 2 units, and yet another may use 2 longs and 12 units. All of these representations are correct. A teacher needs to be open to other ways that children "see" things and then eventually guide students to a final representation using the fewest number of pieces.

B. Discuss with other members of your group how you each recorded your answers.

> **TEACHING NOTE:**
> A teacher should allow students to make the decision as to how their findings are to be recorded: some will write sentences, some will make charts, some will draw pictures.

C. Write the numerals in (b) – (h) as was done in (a).

> **TEACHING NOTE:**
> After children have had many opportunities to develop place value meanings using concrete materials along with oral communication and symbolic notation, they need to make a transition to the pictorial representations along with symbolic notation.

(a)

(b)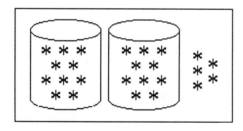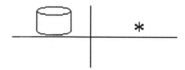

(c)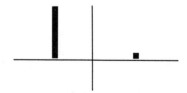

(d)

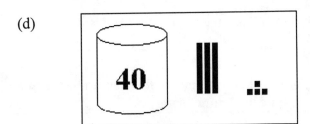

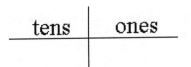

(e)

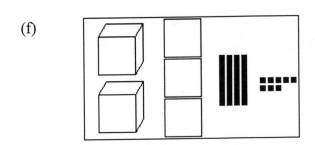

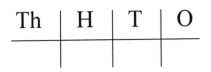

(f)

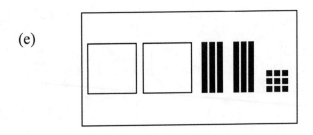

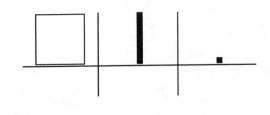

(g)

Th	H	T	O

(h)

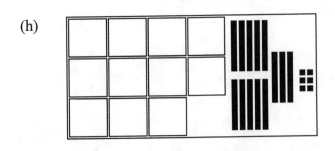

Th	H	T	O

D. In each of the following, you are given either a pictorial model, an oral or word name, or a numeral for a base ten number. Supply the other two in the chart.

	MODEL	ORAL NAME	Numeral in Place Value Table			
			Th	H	T	O
(a)			1	2	5	4
(b)		four hundred thirty-six	Th	H	T	O
(c)		three thousand fifty-four	Th	H	T	O
(d)		one thousand three hundred	Th	H	T	O
(e)		two thousand seven	Th	H	T	O
(f)			Th	H	T	O
(g)			Th	H	T	O

ACTIVITY 4-2: Who Am I?
(Number Sense and Place Value Concepts)

Overview: Students need to use problem-solving and reasoning skills in a variety of situations. This activity focuses on the "use manipulatives" problem-solving strategy. You will discover that there may be more than one "right answer" to a problem.

Materials: Base ten blocks: hundreds (flats), tens (rods), ones (units); clue sheet.

Determine possible solutions using clues and base ten blocks.

(a) I have 4 base ten blocks.
    Some are rods and some are units.
    Their value is less than 20.
    Who am I? _____

(b) I have 7 base ten blocks.
    Some are units and some are rods.
    Their value is an even number between 30 and 40.
    Who am I? _____

(c) I have 4 base ten blocks.
    Some are rods and some are units.
    Their value is between 30 and 50.
    Who am I? _____

(d) I have 5 base ten blocks.
    Some are flats, some are rods, some are units.
    I am a palindrome.
    Who am I? _____

(e) I have 5 base ten blocks.
    Only one of them is a rod.
    Who am I? _____

(f) I have 3 base ten blocks.
    I have no units.
    Who am I? _____

(g) I have 8 base ten blocks.
    There are the same number of flats and units.
    There are fewer rods than flats or units.
    Who am I? _____

(h) I have 16 base ten blocks.
    There is at least one hundred.
    Who am I? _____

Activity 4-3: Roll to One Hundred/ Roll to Zero

Overview: Addition and subtraction are inverses of each other. Addition is the "putting together" of disjoint sets of objects and subtraction can be thought of as "taking away" a subset of objects from a given set. The algorithms for addition and subtraction are step-by-step procedures that can be understood by thinking the process through using base ten materials. In this activity you will use base ten pieces to visually illustrate addition and subtraction with regrouping.

These games should be played in groups of 2-4.

Materials: Base ten pieces, 2 dice, place value mats (found on Appendix Page 3)

(a) **Roll to One Hundred:** A student rolls the two dice and finds the sum of the dots on the faces. That number represented with base ten pieces is put on the student's place value mat. At the end of a player's turn, the pieces should be traded/regrouped to represent the total using the fewest number of pieces. The first player to make a trade to get a flat (100 piece) wins the game. This game builds the concept of addition with regrouping.

Variation: on every fourth turn, the player must subtract the sum of the dice from the total on the mat.

(b) **Roll to Zero:** Each student puts a flat (100 piece) on the place value mat. A student rolls the two dice and finds the sum of the dots on the faces. That number represented with base ten pieces is taken away from the base ten piece/pieces on the place value mat with appropriate trading/regrouping done as needed. The first player to get to zero (clear the mat) wins the game. This game is "harder" to finish as the number of units gets closer to zero. A student may choose to roll just one die instead of the two towards the end of the game. This game builds the concept of subtraction with regrouping.

Activity 4-4: Subtraction Situations

Overview: Teachers must give verbal problems that present a variety of situations requiring subtraction. There are four different situations which students should learn to recognize as subtraction: take-away, comparison, missing addend, and set/subset.
Work with a partner.

Materials: One-inch squares (cut using the Ellison Letter Machine™), chips, or coins.

A. With a partner, use one-inch squares to model and explain each of the situations below that give a verbal problem for the subtraction sentence 6 - 2 = 4.

(a) In take-away subtraction, a set is modeled and a part of it is "taken away." The number of objects remaining represents the difference.
Cassie has six tennis balls and she gives Racine two. How many does Cassie have left?

(b) In comparison subtraction, two disjoint sets are modeled and the objects in each set are compared by matching them one-to-one. The question is asked: How many more in one set than the other? or How many fewer in one set than the other?
Jeff has 6 tapes and Joe has 2 tapes. How many more does Jeff have than Joe? (or How many fewer does Joe have than Jeff?)

(c) In missing addend subtraction, the subtrahend is modeled and the question is asked: How many more are needed to make a total? This is often written symbolically as 2+q =6.
Sarina has $2. She needs $6 to buy a ticket to the ice skating show. How much more does she need to buy a ticket?

(d) In the set/subset model, there is a set of objects. Some of those objects have a specific attribute. The rest of them do not have that attribute. The question is asked: How many in the set do not have the stated attribute?
Mrs. Guenther has 6 students on her Odyssey of the Mind team. Two of them are boys. How many are girls?

TEACHING NOTE:
It is critical that the questions for comparison subtraction be used interchangeably. Young children often *miss* those *how many fewer* questions, not because of a lack of understanding, but because of a lack of familiarity with the wording of the questions.

B. Identify each of the following subtraction situations as take-away, comparison, missing addend, or set/subset. How could manipulatives be used to model each situation?

(a) Mrs. Ouzts has 25 students in her *Transition Mathematics* class and 14 of them are girls. How many boys are in her class?

(b) Teri wants to collect all 18 of the trading cards of the Atlanta Braves. If she now has 8 cards, how many more does she need to collect?

(c) Eli has 8 pencils. If he gives 3 to his sister, how many will he have then?

(d) Mrs. Spero has 11 jars but only 8 lids. How many jars will not have lids?

(e) If Jerry buys 4 more notebooks, he'll have a dozen. How many notebooks does he have now?

(f) Duchess gave birth to 8 puppies. Five of them were female. How many male puppies were there?

(g) Kendle wrote 11 letters to her Space Academy friends. Kate wrote only 3. How many fewer letters did Kate write than Kendle?

(h) Mr. Snope has a group of seventh graders on his MathCounts team. If he gets 5 eighth graders to join the group, he'll have 12 students on the team. How many seventh graders are on the team?

TEACHING NOTE:

As so much of mathematics is symbolic, students need to be able to interpret what symbols mean, especially when they are used together. Juxtaposing, or concatenating, symbols has different meanings in mathematics, depending on the context. For example, juxtaposing the numerals 2 and 3 deals with numeration/place value: 23 is twenty-three. Juxtaposing 2 with the variable a means multiplication: $2a$. Juxtaposing 2 and $\frac{1}{3}$ means addition: $2\frac{1}{3} = 2 + \frac{1}{3}$.

On the other hand, if A and B represent points, then juxtaposing A and B means the length of the segment from A to B (and not multiplication as some students may interpret). AB = 5.

Activity 4-5: The Difference Is ...

Overview: This activity helps students develop understanding of subtraction with regrouping using different models. It is written for 2-digit subtraction but can be extended to 3-digit subtraction.
Students should work in pairs and discuss their work.

Materials: Place value mats (found on Appendix Page 3), base ten pieces, and place value dice [Place value dice have numerals instead of dots. The ones die and the tens die differ in color or size.]

A. For the take-away model of subtraction (the model for the standard subtraction algorithm), each student rolls a set of place value dice. The student rolling the larger number builds it on the place value mat using base ten pieces. The student rolling the smaller number takes away that number represented using base ten pieces, making a fair trade, if necessary. In the event that both students roll the same number on the dice, each student rolls again. Perform this activity three times.

B. For the comparison model of subtraction, each student rolls a pair of place value dice and models the number using base ten pieces. Students compare their numbers by matching their pieces one-to-one (making fair trades if necessary). The pieces that are not "matched" represent the difference between the two numbers. Of course, if each student rolls the same number, the two sets of base ten pieces are in a one-to-one correspondence, and the difference is zero. Perform this activity three times.

Activity 4-6: Make Four Flats

Overview: This activity allows students to solve missing addend problems concretely. They begin by modeling a number with base ten pieces and then determining how many more pieces they would need to make four flats (400). It is important for each student to work with a partner so each can explain the processes used in solving the problem. This work with concrete materials allows students to experience a concept in a different way, even though they may already know algorithmic procedures.

Work with a partner.

Materials: Base ten pieces

BUILD THE NUMBER	HOW MUCH MORE TO MAKE 4 FLATS?
142	
299	
386	
207	
47	
28	

Students who understand the structure of numbers and the relationships among numbers can work with them flexibly (Fuson 1992). They recognize and can generate equivalent representations for the same number. For example, 36 can be thought of as 30 + 6, 20 + 16, 94, 40 – 4, three dozen, or the square of 6. Each form is useful for a particular situation. Thinking of 36 as 30 + 6 may be useful when multiplying by 36, whereas thinking of it as 6 sixes or 9 fours is helpful when considering equal shares. Students need to have many experiences decomposing and composing numbers in order to solve problems flexibly.

- Principles and Standards for School Mathematics

Activity 4-7: "Fair Trades" and the Trading Game

Overview: The concept of "fair trade" is critical as students regroup in all four whole number operations. In addition to reinforcing this critical skill, this activity could also serve as a stepping stone to a unit in other number bases.
   This game is for 2-5 players.

Materials: Yellow, blue, red, and green chips or squares; one die; a trading mat found on Appendix Page 4.

• 3 yellow pieces can be traded for a blue piece.
• 3 blue pieces can be traded for a red piece.
• 3 red pieces can be traded for a green piece.

Player One rolls the die and takes that number of yellow pieces. The player must trade if possible. For example, if a player rolls a "four," he puts four yellow pieces on his mat. Then he trades three of the yellow pieces for one blue piece. After no more trades can be made, the next player takes a turn. Players take turns rolling the die, putting on yellow pieces, and making possible trades. The first player to trade for a green piece wins the game. Each player keeps track of his plays on a sheet like the one below.

# green	# red	# blue	# yellow	# on die this turn	Cumulative sum of die

Study the relationship between the numbers of different colored pieces in each row and the cumulative sum of the rolls of the die for that row. Describe the pattern that you find. What is the "value" of each colored piece?
Variation: Use uneven place values for the colored pieces. For example, let three yellows be traded for one blue, eight blues for one red, and two reds for one green. This relates to measurements as in the U.S. Customary system.

G	R	B	Y

Activity 4-8: Multiplication Matters (I)

Overview: In this activity we will focus on multiplication as repeated addition. In Activities 4-9 and 4-10 we will model multiplication using the Cartesian product and using the rectangular array area model.
Work in pairs.

Materials: For Part A: Counters such as teddy bears, 1-inch squares, chips, buttons. For Part C: Place value mat (Appendix Page 3) and base ten pieces.

A. **Multiplication as Repeated Addition:**

(a) In your workspace, place two groups of 5 chips. If these two groups are "pushed together," what operation does this model?
_____ How many chips are there altogether? _____

Write an addition sentence that means "two groups of 5 is 10."

(b) In your workspace, place three groups of 4 squares. If these three groups are "pushed together," what operation is this?
_____ How many squares are there altogether? _____

Write an addition sentence that means "three groups of 4 is 12."

(c) Make up two problems like the ones above and ask a partner to model them and write an addition sentence for each one.

TEACHING NOTE:
In presenting this material to children, the teacher should model each problem on the overhead (after the children) or have a child come to the overhead projector to model and explain.

(d) Write a verbal statement and an addition sentence indicated by the following picture:

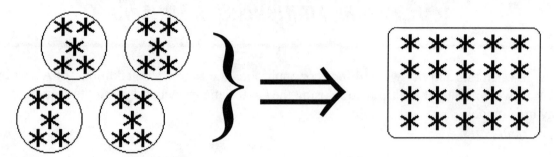

B. Consider the two addition problems below (not the computational answer)

$$7 + 1 + 3 + 4 + 2 \qquad \text{and} \qquad 8 + 8 + 8 + 8 + 8$$

(a) How are they alike?

(b) How are they different?

(c) $8 + 8 + 8 + 8 + 8$ is a repeated addition problem. The addend 8 is used five times.
Mathematicians use a shortcut notation for repeated addition and would write:
$8 + 8 + 8 + 8 + 8 = 5 \times 8$. 5×8 means "5 groups of 8" and we read 5×8 as "5 times 8."
This operation is called multiplication. The first number in the multiplication problem
tells us how many groups and the second number tells us how many or how much is in
each group.

Write each multiplication problem as an addition problem:

$5 \times 2 =$

$3 \times 10 =$

$7 \times 4 =$

(d) Write each addition problem as a multiplication problem:

$6 + 6 + 6 + 6 =$

$9 + 9 + 9 + 9 + 9 + 9 =$

$12 + 12 + 12 =$

TEACHING NOTE:

2×5 and 5×2 are not modeled the same way.

$\qquad 2 \times 5 = 5 + 5$ (2 groups of 5)   and $5 \times 2 = 2 + 2 + 2 + 2 + 2$ (5 groups of 2)

Even though the computational result (10) is the same, the conceptual models for the two
problems are different.

C. Explore multiplication (repeated addition) with base ten materials.

(a) Using ones and tens and place value mats, show the following and determine the answer represented. Observe that these representations require no regrouping.

2 tens

4 twenties

3 thirteens

(b) Again using base ten pieces, show the following. Use fair trading to find out how much each is altogether. After each representation, discuss with your partner how you determined this product.

5 thirteens

3 twenty-fours

5 sixteens

(c) Use base ten pieces to show the following. At each step, discuss your thought processes with your partner.

2 groups of one hundred thirty-four

4 groups of one hundred twenty-seven

(d) Look at this picture and combine the base ten pieces and record the product:_____

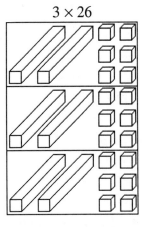

3×26

TEACHING NOTE:

In working with children on this and related activities, the following guidelines are important:
  (i) Work first with numbers that can be represented with ones and tens. The initial problems should not require regrouping.
  (ii) When presented with the first problem that requires regrouping, review the regrouping process with the children.
  (iii) As students work, pause periodically and require that they verbally describe the way they are thinking.
  (iv) After the students have become accomplished with numbers represented using ones, tens, and hundreds pieces, ask them to work on problems represented pictorially.

Activity 4-9: Multiplication Matters (II)

Overview: One other definition of multiplication uses a Cartesian product approach. Let a and b be any whole numbers. If $n(A) = a$ and $n(B) = b$, then $ab = n(A \times B)$. For example, if $A = \{p, q\}$ and $B = \{x, y, z\}$, then $A \times B = \{(p, x), (p, y), (p, z), (q, x), (q, y), (q, z)\}$. Note that $n(A) = 2$, $n(B) = 3$, and $n(A \times B) = 2 \times 3 = 6$. This is a fairly abstract development and will not be used quite in this form with younger students. Multiplication as a Cartesian product can be developed by using food – a great motivator! After solving a problem at the concrete level, students should move to the pictorial and then abstract levels.

Materials: Two kinds of crackers (Ritz, Saltine) and three kinds of toppings (cheese, peanut butter, and jelly).

A. **Multiplication as a Cartesian Product:**

   (a) Work with a partner. With the crackers and toppings, make all the different snacks that can be made using one cracker and one topping. (This is problem solving at the concrete level.)

   (b) Work individually or with your partner. Draw pictures to show the solution to the problem. Be sure there is a legend showing what the different symbols mean. (This is problem solving at the pictorial level.)

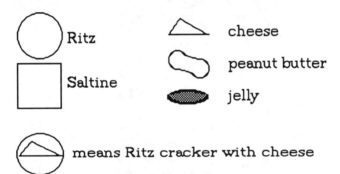

   (c) Write the solution to the problem as a set of ordered pairs of the form (cracker, topping). If $A = \{$Ritz, Saltine$\}$ and $B = \{$cheese, peanut butter, jelly$\}$, determine $A \times B$. (This is problem solving at the abstract level.)

   $A \times B =$ _____

TEACHING NOTE:

Another way to model multiplication is with a number line. The number line is a semiconcrete model and is a convenient way to "visualize" arithmetic operations. 3×4 would be modeled as 3 "jumps" of 4 units on the number line. So, $3 \times 4 = 12$.

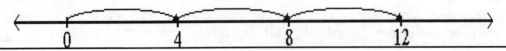

Activity 4-10: Multiplication Matters (III)

Overview: The rectangular array area model of multiplication is critical in helping students connect a multiplication model to multiplication algorithms, to the distributive property, and later to algebra. One bonus is the connection to geometry and measurement. The real bonus is that choosing this model for multiplication will support learning at other levels.

Materials: One-inch squares of assorted colors (can be cut with Ellison Letter Machine™) or colored square tiles; base ten pieces; square grid paper (Appendix Page 6)

A. Using squares, make a rectangle with two rows of three squares, matching the squares edge-to-edge. This is called a "two-by-three" rectangle or a 2 × 3 rectangle. Note that the area of the rectangle is 6 square inches (or 6 square units), since area is a measure of covering. The area (6) of the rectangle is the product of its two dimensions. We also say that 2 and 3 are factors of 6.

We can model 3 × 5 as a rectangle made up of three rows of five squares. Its area is 15 square units, or the product of its two dimensions.

   (a)   Using squares, demonstrate 4 × 6 as a rectangular array area model.

   (b)   Using squares, demonstrate 6 × 4 as a rectangular array area model.

   (c)   How are the two models alike?

   (d)   How are they different?

   (e)   Using square grid paper, cut out a 4 × 6 rectangle and a 6 × 4 rectangle. What word geometrically describes how the two rectangles are related?

Since the 4 × 6 rectangle and the 6 × 4 rectangle are congruent rectangles, they are not distinct geometrically. Modeling the concept of multiplication, however, one has 4 rows of 6 and the other has 6 rows of 4; and they are "different." Since they have the same area (24 u^2), we can say that the products of the two dimensions are the same. That is, 4 × 6 = 6 × 4. This illustrates the *commutative property of multiplication*.

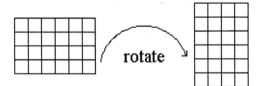

rotate

B. Using base ten pieces with the small square having dimensions 1 unit × 1 unit, what are the dimensions of
(a) the "long" or tens rod?

(b) the "flat" or hundreds piece?

(c) Using the base <u>ten</u> pieces, 3 × 12 can be modeled as a rectangle with dimensions of 3 units and 12 units, or as a 3 × 12 rectangle. We consider 12 to be a "binomial" as it is modeled with 1 ten and 2 ones—two different base ten pieces, or "unlike terms." (Note: The expression $ax + b$ is a binomial of degree 1 in x. Thus, $12 = 1 \times ten + 2$ is a binomial in *ten*.)

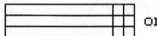

This geometric representation is related to the partial product algorithm below.

		t	o
		1	2
	×		3
			6	3 ones × 2 ones
	+	3	0	3 ones × 1 ten
		3	6

3 tens   and 3 twos

30   +   6   = 36

t	o

Model 2 × 24 as a rectangle. Draw a geometric representation and relate it to the partial product algorithm.

TEACHING NOTE:

When the product 3 × 12 is written vertically as $\begin{smallmatrix} 12 \\ \times\ 3 \end{smallmatrix}$, we call 3 the multiplier. 3 is the second (lower) number in the product, but the product is read "3 times 12."

(d) What are the dimensions of the rectangle below?   height: _____   base: _____

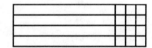

What product gives the area of this rectangle? _____ × _____

When the pieces are rearranged (and fair trades are made where needed), what number represents this area? _____

Show how the model in (d) on the previous page is connected to the partial product algorithm.

(e) Model the following simple multiplications with base ten pieces and then use the insight gained to compute them mentally.

Example: 2×13 $\Rightarrow$

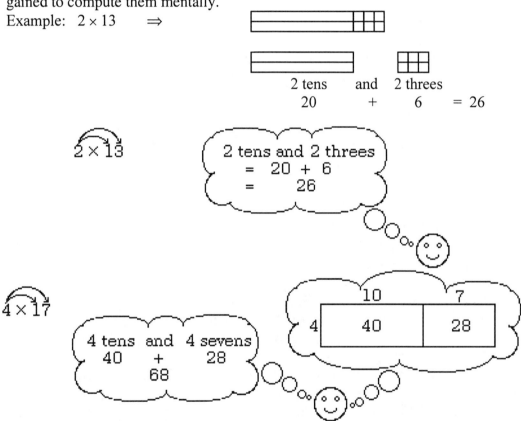

2 tens   and   2 threes
20   +   6   = 26

Compute the following products using base ten pieces and then mentally:

(i) $3 \times 14 =$ (ii) $4 \times 15 =$ (iii) $4 \times 18 =$

TEACHING NOTE:
The problems in this activity are practical applications of the distributive property. It is critical that students "practice" this property numerically in a meaningful way before seeing it represented abstractly as $a(b + c) = ab + ac$.

(f) To model 17×12, we would build a 17×12 rectangle with base ten pieces:

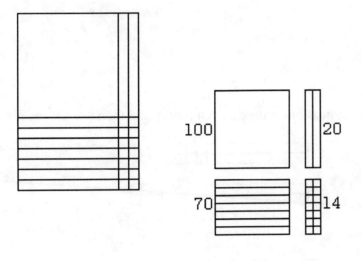

If we separated the tens and ones in each factor, we could separate these pieces into four rectangles with areas as shown. These areas are the four partial products in the multiplication algorithm.

```
  h  t  o
     1  2
  ×  1  7
     1  4    7 ones × 2 ones
     7  0    7 ones × 1 ten
     2  0    1 ten × 2 ones
 +1  0  0    1 ten × 1 ten
  2  0  4
```

We could "picture" 23×12 by separating the tens and ones in each factor, calculating the area of the four rectangles, and adding these four areas to give the area of the 23×12 rectangle (the product of its 2 dimensions).

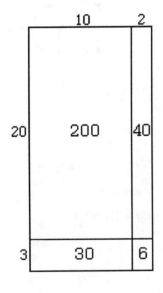

The four areas are connected to the four partial products in the multiplication algorithm.

```
  200
   40
   30
 +  6
  276
```

```
  h  t  o
     1  2
  ×  2  3
        6    3 ones × 2 ones
     3  0    3 ones × 1 ten
     4  0    2 tens × 2 ones
 +2  0  0    2 tens × 1 ten
  2  7  6
```

Follow examples in the previous sections to build rectangles to compute the following products. Connect the model to the partial product algorithm.

5×27 26×13

TEACHING NOTE:

This rectangular array area model with its related partial product algorithm will reappear in multiplying fractions (Activity 6-16), in multiplying decimals (6-19), in multiplying a natural number and a mixed number, and also in algebra. For example, to compute $3 \times 2\frac{1}{3}$, we draw a rectangle whose dimensions are 3 and $2\frac{1}{3}$. We separate $2\frac{1}{3}$ into its component parts, calculate the areas of the two rectangles, and add them together.

$$3 \times 2\tfrac{1}{3}$$

$$3 \times 2\tfrac{1}{3} = \text{three 2's} + \text{three } \tfrac{1}{3}\text{'s}$$

$$= 6 + 1 = 7$$

To find $a(a + 5)$ using the area model, we draw a rectangle whose dimensions are a and $a + 5$. Since $a + 5$ is a binomial, we separate the a and the 5, calculate the areas of the two rectangles, and add them to get the area of the $a \times (a + 5)$ rectangle.

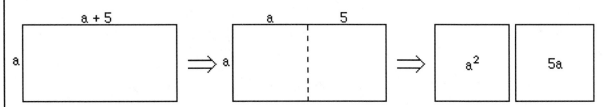

So, $a(a + 5) = a^2 + 5a$. Similarly, $(x + 2)(x + 3)$ is the area of a rectangle with the dimensions $x + 2$ and $x + 3$.

The areas of the four rectangles are the four partial products found by multiplying the binomials: $(x + 2)(x + 3) = x^2 + 2x + 3x + 6 = x^2 + 5x + 6$.

People who never get carried away should be.
- Malcolm Forbes

Activity 4-11: Divvy Up—Fair Shares

Overview: Although the concept of division comes naturally to students as they learn to share, the long division algorithm is typically difficult to master. Without base ten manipulatives, students have problems connecting the concept of division to the algorithm (the use of abstract symbols). There are two basic concepts of division: partition division and measurement division. The one you probably think of most often is that of separating (partitioning or dividing) a set of objects into some number of parts, each having the same amount.

For example: If 15 children are divided into 3 teams (parts), how many children will be on each team?

This concept of division is referred to as *partition division* (or partitive division), because we know how many equal-sized parts into which the set is divided. We are asked to find the number of objects or things there will be in each of the parts. This division is one of equal sharing or *divvy up—fair shares*. Partition division is the model for the long division algorithm.

The second concept of division is called *measurement division* and can be modeled by repeated subtraction. Multiplication can be thought of as repeated addition; thus, its inverse operation, division, can be thought of as repeated subtraction. A typical question: How many 7's are in 21? A good way to help children understand that measurement division is repeated subtraction is to bring in a 5-pound bag of sugar, a 1-cup measuring cup, and a bowl. Ask: How many cups of sugar are in a 5-pound bag? Have students first estimate and discuss how they determined their estimates. Then have a student actually open the bag, measure out (subtract) one cup and continue until the sugar in the bag is depleted.

For example: If 20 jawbreakers are to be put into bags with 5 jawbreakers in each bag, how many bags are used?

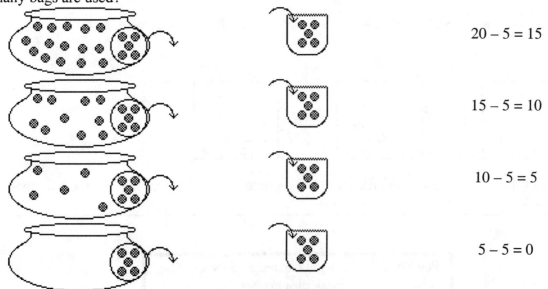

$$20 - 5 = 15$$

$$15 - 5 = 10$$

$$10 - 5 = 5$$

$$5 - 5 = 0$$

So, four groups of jawbreakers were "measured" out or taken out. $20 \div 5 = 4$. (Note: alternately, a student might count by fives from 0 up to 20, saying "5, 10, 15, 20" while keeping track that he or she has used 4 fives.)

Either the partition model or the measurement model can be used to understand the long division algorithm. In this activity we will look at the long division algorithm in terms of the partition model.

Work in groups of four.

Materials: Base ten pieces for each student (commercial, or those cut with the Ellison Letter Machine™ and the three base ten dies) and place value mat (Appendix Page 3)

A. **Partition Division Activity:**
   (a) In your group discuss the "divvy up" process by explaining how children go about trying to "divvy up" candy, etc., at home with brothers/sisters or between friends. (Hint: "one for you, one for me" routine) Discuss what is meant by "fair shares."

   (b) Count out 9 of something and have a group member show how to "divvy up" among 3 students. Relate this to $9 \div 3 = 3$.

   (c) Divvy up 3 packs of gum and 6 extra pieces among three students so that each student has a fair share. How much is a fair share of gum?

   (d) Using your place value mat and base ten materials, show a representation of 26 in the most efficient way on your mat. Divide 26 into 2 equal-sized groups on the mats. Continue with examples requiring no regrouping, such as 39 split 3 ways, or 44 split 4 ways, or 28 split 2 ways. One group member should record the process to show that long division is simply a written record, in the language of mathematics, of what is being done physically with ones and tens.

   (e) Model 35 using base ten blocks. Divide 35 into 3 equal-sized groups. How many are in each group? _____ What is different about this problem from the ones before?

   (f) Prepare for regrouping in division by displaying 24 and then divvy up 3 ways. (Remember, you have the materials necessary to solve the problem and you must have a total of 24 on your mats at all times.)

   (g) Once you can explain what you are doing with division problems similar to the ones above, divvy up the following 3-digit numbers into groups of three. Notice that the first two can be computed without regrouping and that the last two have remainders. Discuss each problem with your group members.

363	609
213	162
221	176

B. **Connecting the Partition Division Activity to the Algorithm:** Have one member of your group model 679 on a table. Group members should then divvy up 679 into five equal shares. Be sure to take this opportunity to use the division vocabulary: dividend, divisor, quotient, remainder. Now connect the division with the algorithm. Record the steps symbolically. Write about the process using words in sentences.

(a) Into how many groups are you dividing 679? _____

(b) How many hundreds are in each group? _____

(c) How much did you take away from 679? _____

(d) What do you have left? _____

```
   H T O
     1
5 )6 7 9
  -5 0 0
   1 7 9
```

(e) To divvy up the tens, do you have to make any fair trades before dividing up the tens? _____ How many tens will each group get? _____ Record 3 in the tens place.

(f) How much did you take away from the 179?_____ What do you have left?_____

```
   H T O
     1 3
5 )6 7 9
  -5 0 0
   1 7 9
  -1 5 0
     2 9
```

(g) To divvy up the ones, do you have to make any fair trades before dividing them among the five groups? _____ How many ones will each group get? _____ Record in the ones place.

(h) How much did you take away from the 29? _____ What do you have remaining?_____

(i) In this problem, what is the dividend? _____ what is the divisor? _____ what is the quotient? _____ what is the remainder? _____

```
   H T O
     1 3 5
5 )6 7 9
  -5 0 0
   1 7 9
  -1 5 0
     2 9
    -2 5
       4
```

Now carefully analyze the problem of dividing 397 into 6 equal parts. Model the division with base ten pieces and then use the algorithm answering the questions you discussed in (a)-(i).

C. **Division Situations:** Finding a missing factor is a division situation. Recall that if $a \times b = c$, the multiplier a tells the number of groups and the multiplicand b tells the number of objects in each of the groups.

- $4 \times ? = 12$ is a *partitive division* situation since we know there are 4 groups. The question asked is: If 12 is divided into 4 groups, how many are in each group?

- $? \times 5 = 30$ is a *measurement division* situation since we know there are 5 things in each group and we are looking for the number of groups. The question asked is: How many groups of 5 are in 30?

Identify each of the following verbal problems as measurement or partition division and then determine the answer.

(a) Bottles are packed 24 to a case. How many cases are needed to pack 480 bottles?

(b) A teacher calculates the test average for Juanita who made 92, 88, and 84 on three tests. What is Juanita's average?

(c) How many dimes are in a dollar?

(d) Cassie shares a bag of 36 "Mary Jane's" with 3 of her friends. How many pieces of candy will each person get?

(e) Two hundred eighty-eight students at DeRenne Middle School are to be divided into nine homerooms. How many students will be in each homeroom?

(f) Jamal has 24 pencils. He stands at the classroom door and gives 2 pencils to each student who comes into the class. How many students get pencils?

(g) Eggs are sold by the dozen. How many dozen eggs are in a basket with 72 eggs?

D. **Sensible Division Estimates:** As number sense and the ability to mentally estimate are required in long division, students need formal experience in making sensible estimates.

For Example: The total attendance for 6 Atlanta Braves games in the Fulton County Stadium was 19,436. The average attendance per game was about: (?)   300   3,000   30,000

$6\overline{)19436}$ Are there enough ten thousands to divide? $6\overline{)19///}$

Are there enough thousands to divide? $6\overline{)19///}$

The quotient is in the thousands and it has four digits. $6\overline{)19///}$
So, 3,000 is a sensible estimate.

Make "marks" to show how many digits are in the quotient. Choose a sensible estimate.

(a) 175 students ride on 4 buses. $4\overline{)175}$

   There are about   4   40   400   students per bus.

(b) You have 36 months to pay off a loan of $7,150. $36\overline{)7150}$

   Payments are about $20 $200 $2,000   per month.

TEACHING NOTE:

In developing the long division algorithm, students must have time to explore division with base ten pieces to develop the concept concretely; some will "invent" their own ways to write symbolically what they have done physically. It is the job of the teacher to connect the concrete experiences with the abstract symbolism of the long division algorithm and to encourage students to visualize the materials as they try the written algorithm. You may ask yourself: If the long division algorithm is the partition division model, isn't it enough to just focus on partition division? NO! It is critical for you to develop *both* division concepts with children. The notion of grouping is fundamental to understanding place value. "How many groups of 10?" is asked in place value; this is measurement division. Later, division with fractions will require both partition and measurement division. For example, $\frac{6}{7} \div 2$ is modeled by partition division.

"If $\frac{6}{7}$ is divided into 2 equal-sized parts, how much is in each part?"

On the other hand, $\frac{8}{3} \div \frac{2}{3}$ must be modeled by measurement division:

"How many groups of $\frac{2}{3}$'s are in $\frac{8}{3}$'s?"

Activity 4-12: Whole Number Operations with Cuisenaire® Rods

Overview: Cuisenaire® Rods can be used to develop concepts of whole numbers (and fractions, geometry, measurement). Each rod has a square centimeter base, and the rods vary in length from one to 10 centimeter. The rods have codes which are alphabet letters: **w**hite, **r**ed, **g**reen, **p**urple, **y**ellow, **d**ark green, blac**k**, brow**n**, blu**e**, **o**range. Students need time to become familiar with colors and sizes of rods as they begin considering the relationships among them. One advantage to using the rods is the fact that they are metricized, so students can become familiar with metric lengths from 1 to 10 centimeters. Students can also make their own "rods" by coloring centimeter strips or by taping together small construction paper squares. When two rods are put end-to-end, we call it a "train."

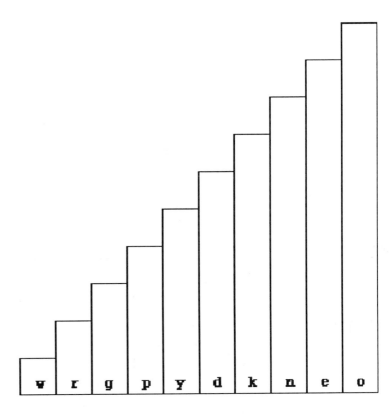

Materials: Cuisenaire® Rods or "rods" made from centimeter strips or paper squares.

A. **Addition:** Take an orange rod (10 cm). Below are two different 2-rod trains which have the same length as the orange rod.

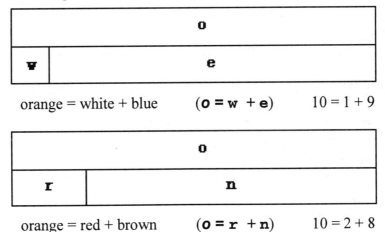

orange = white + blue   (**o = w + e**)   10 = 1 + 9

orange = red + brown   (**o = r + n**)   10 = 2 + 8

(a) Find all different 2-rod trains which have the same length as the orange rod.

(b) Find all different 3-rod trains which have the same length as the brown rod.

B. **Subtraction:**

(a) *missing addend* : What rod must be used with the green rod to make a train as long as the blue rod? **g + ? = e**

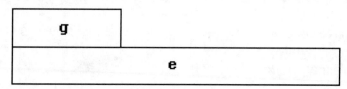

(b) *comparison*:: How much longer is the black rod than the red rod? **k − r = ?**

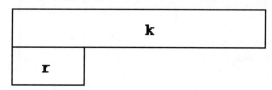

C. **Multiplication:** What rod is as long as a train made from 3 green rods? **3 × g = ?**

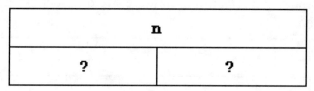

D. **Division:**

(a) *Partitive Division*:: Two rods of the same color make a train as long as the brown rod. What color are they? **n ÷ 2 = 2**   or   **2 × ? = n**

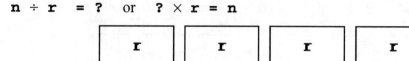

(b) *Measurement Division*:: How many red rods make a train as long as a brown rod?
n ÷ r = ?   or   **? × r = n**

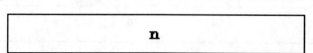

Activity 4-13: Value This

Overview: A calculator is a necessary and useful tool in helping students develop number sense. It allows for trial and error in the testing of conjectures and avoids the tedium of pencil-and-paper computations

Materials: A calculator

Using a calculator and the digits 2, 3, 5, 8, and 9, find the largest and smallest possible value.

<u>largest value</u>

$\square\square + \square\square\square$

$\square\square \times \square\square\square$

$\square\square\square\square \times \square$

$\square\square\square\square - \square$

<u>smallest value</u>

$\square\square + \square\square\square$

$\square\square \times \square\square\square$

$\square\square\square\square \times \square$

$\square\square\square\square - \square$

TEACHING NOTE:
Students need to interpret symbolic expressions in various ways. For example, **a + 3** is read "a plus **3**" but it means
* The sum of **a** and **3**,
* **3** more than **a**,
* **3** added to **a**, and
* **a** increased by **3**.

(See Activity 4-14.)

Activity 4-14: I Have . . . Who Has?

Overview: This activity is for an entire class of students. Its purpose is to foster the language of whole number operations and to assist in the development of mental computation skills. Cards are distributed and any student may begin reading his/her card to start the round. (The round will be over when it gets back to the student who began.)

Materials: Index cards and a xeroxed copy of this page with each "I have . . . Who has?" strip glued to an index card.

I have 63. Who has that divided by 9?

I have 7. Who has that plus 3?

I have 10. Who has that minus 7?

I have 3. Who has that multiplied by 6?

I have 18. Who has 12 more than that?

I have 30. Who has that divided by 6?

I have 5. Who has 8 times that?

I have 40. Who has 4 less than that?

I have 36. Who has that divided by 9?

I have 4. Who has the sum of 9 and that number?

I have 13. Who has one less than twice that number?

I have 25. Who has that times 3?

I have 75. Who has that decreased by 3?

I have 72. Who has that divided by 9?

I have 8. Who has that increased by 9?

I have 17. Who has twice that?

I have 34. Who has one less than that?

I have 33. Who has that divided by 3?

I have 11. Who has 9 more than that?

I have 20. Who has 3 times that?

I have 60. Who has that minus 15?

I have 45. Who has that divided by 5?

I have 9. Who has that multiplied by 9?

I have 81. Who has that plus 9?

I have 90. Who has that decreased by 20?

I have 70. Who has that minus 7?

TEACHING NOTES: A SUMMARY

We have to rethink how computation is accomplished today in light of technology, as complex computation is usually done with calculators and computers. Students need to know when to choose among computational strategies: mental computations, estimation, pencil-and-paper, or calculators/computers. Teachers need to have reasonable expectations for proficiency with pencil-and-paper computation. Clearly, pencil-and-paper computation cannot continue to dominate the curriculum, or there will be insufficient time for children to learn other, more important mathematics they need now and in the future. It is the job of teachers to help children master basic facts and algorithms and understand their usefulness and relevance to daily situations. This task is accomplished by emphasizing underlying concepts, using physical materials to model procedures, linking the manipulation of materials to the steps of the procedures, and developing thinking patterns.

After mastering the concepts of the four operations, students should explore operation sense—the investigation of relationships among the operations as well as the investigation of properties of an operation. Operation sense also involves the learning about the effects of an operation on two numbers. For example, adding 20 to a whole number yields a result much smaller than multiplying the number by 20 and much larger than dividing the number by 20.

Students also need to develop oral language to accompany the four basic operations. They should recognize that "3 + 4" can be read "three plus four," but that it also means "the sum of 3 and 4" or "4 more than 3" or "3 increased by 4." Many students only recognize "10 - 3" as "ten minus three," but they should also recognize it as "3 subtracted from 10" or "10 decreased by 3" or "3 less than 10." "Total," "sum", "how many in all?" and "how many altogether?" usually refer to addition. Students should not just look for key words; they should learn to analyze a situation. For example: *Mandy now has four pens. She had a total of nine pens at the beginning of her Language Arts class. How many in all were loaned to her friends?*

When students take a dozen objects and divide them into two equal-sized groups, they should see each of those groups, also, as $\frac{1}{2}$ of twelve. The third grade teacher who introduces the symbols for division needs to make sure that by the time students finish the third grade, they are able to symbolically represent "twelve divided by two" in four ways: $12 \div 2$, $2\overline{)12}$, and $\frac{12}{2}$.

They should also recognize that it means the same as $\frac{1}{2}$ of 12. The latter two representations are needed for fraction situations.

As we study whole numbers, then integers, and then rational numbers in Chapters 4, 5, and 6, we need to keep in mind the activities and concepts that are appropriate to each grade level. Review the following comments from the narrative portion of the NCTM *Principles and Standards for School Mathematics.*

In grades 3–5, students' study and use of numbers should be extended to include larger numbers, fractions, and decimals. They need to develop strategies for judging the relative sizes of numbers. They should understand more deeply the multiplicative nature of the number system, including the structure of 786 as 7×100 plus 8×10 plus 6×1. They should also learn about the position of this number in the base-ten number system and its relationship to benchmarks such as 500, 750, 800, and 1000.

A significant amount of instructional time should be devoted to rational numbers in grades 3–5. The focus should be on developing students' conceptual understanding of fractions and decimals—what they are, how they are represented, and how they are related to whole numbers—rather than on developing computational fluency with rational numbers. Fluency in rational-number computation will be a major focus of grades 6–8.

As students move from third to fifth grade, they should consolidate and practice a small number of computational algorithms for addition, subtraction, multiplication, and division that they understand well and can use routinely. Many students enter grade 3 with methods for adding and subtracting numbers. In grades 3–5 they should extend these methods to adding and subtracting larger numbers and learn to record their work systematically and clearly. ... Although the expectation is that students develop fluency in computing with whole numbers, frequently they should use calculators to solve complex computations involving large numbers or as part of an extended problem.

In the middle grades, students should continue to work with whole numbers in a variety of problem-solving settings. They should develop a sense of the magnitude of very large numbers (millions and billions) and become proficient at reading and representing them. For example, they should recognize and represent 2 300 000 000 as 2.3×10^9 in scientific notation and also as 2.3 billion. ... Students also need to understand various forms of notation and recognize, for instance, that the number 2.5×10^{11} might appear on a calculator as 2.5E11 or in some other format depending on the make and model of the machine. Students' experiences in working with very large numbers and in using the idea of orders of magnitude will also help build their facility with proportionality.

RESOURCES

Periodicals:

Baroody, Arthur J. "The Development of Basic Counting, Number, and Arithmetic Knowledge among Children Classified as Mentally Handicapped." In *International Review of Research in Mental Retardation,* edited by Laraine Masters Glidden, pp. 51–103. New York: Academic Press, 1999.

Bartek, Mary Marron. "Hands-On Addition and Subtraction with the Three Pigs." *Teaching Children Mathematics* 4 (October 1997): 68-71.

Bird, Elliott. "Counting Attribute Blocks: Constructing Meaning for the Multiplication Principle." *Mathematics Teaching in the Middle School* 5 (May 2000): 568-573.

Brahier, Daniel J. and William R. Speer. "Count on Mathematics for Number Sense." *Teaching Children Mathematics* 2 (February 1996): 351-56.

Caliandro, Christine Koller. "Children's Inventions for Multidigit Multiplication and Division." *Teaching Children Mathematics* 6 (February 2000): 420-424, 426.

Carroll, William M., and Andrew C. Isaacs. "Strategies for Basic-Facts Instruction." *Teaching Children Mathematics* 5 (May 1999): 508-515.

Cobb, Paul, and Grayson Wheatley. "Children's Initial Understandings of Ten." *Focus on Learning Problems in Mathematics* 10, no. 3 (1988): 1–28.

"Computation versus Number Sense." *Mathematics Teaching in the Middle School* 4 (October 1998): 110-112.

Cotter, Joan A., "Using Language and Visualization to Teach Place Value." *Teaching Children Mathematics* 7 (October 2000): 108-114.

Curcio, Frances R., and Sydney L. Schwartz. "There Are No Algorithms for Teaching Algorithms." *Teaching Children Mathematics* 5 (September 1998): 26-30.

"Developing Number Sense: What Can Other Cultures Tell Us?" *Teaching Children Mathematics* 7 (February 2001): 312-319.

"Developing Numbers and Operations with Affordable Handheld Technology." *Teaching Children Mathematics* 7 (November 2000): 162-164.

Fuson, Karen C., Laura Grandau, and Patricia A. Sugiyama. "Achievable Numerical Understandings for All Young Children." *Teaching Children Mathematics* 7 (May 2001): 522-526.

Fuson, Karen C. "Research on Learning and Teaching Addition and Subtraction of Whole Numbers." In *Analysis of Arithmetic for Mathematics Teaching,* edited by Gaea Leinhardt, Ralph Putnam, and Rosemary A. Hattrup, pp. 53–188. Hillsdale, N.J.: Lawrence Erlbaum Associates, 1992.

Fuson, Karen C., Diana Wearne, James C. Hiebert, Hanlie G. Murray, Pieter G. Human, Alwyn I. Olivier, Thomas P. Carpenter, and Elizabeth Fennema. "Children's Conceptual Structures for Multidigit Numbers and Methods of Multidigit Addition and Subtraction." *Journal for Research in Mathematics Education* 28 (March 1997): 130–62.

Graeber, Anna O., and Patricia F. Campbell. "Misconceptions about Multiplication and Division." *Arithmetic Teacher* 40 (March 1993): 408–11.

Graeber, Anna O., and Elaine Tanenhaus. "Multiplication and Division: From Whole Numbers to Rational Numbers." In *Research Ideas for the Classroom, Middle Grades Mathematics,* National Council of Teachers of Mathematics Research Interpretation Project, edited by Douglas T. Owens, pp. 99–117. New York: Macmillan Publishing Co., 1993.

Greer, Brian. "Multiplication and Division as Models of Situations." In *Handbook of Research on Mathematics Teaching and Learning,* edited by Douglas A. Grouws, pp. 276–95. New York: Macmillan Publishing Co., 1992.

"How Huge is a Hundred?" *Teaching Children Mathematics* 6 (November 1999): 154-159.

Isaacs, Andrew C., and William M. Carroll. "Strategies for Basic-Facts Instruction." *Teaching Children Mathematics* 5 (May 1999): 508–15.

Johnson, Aostre N., and Jody Kenny Willis. "Multiply With MI: Using Multiple Intelligences to Master Multiplication." *Teaching Children Mathematics* 7 (January 2001): 260-267, 269.

Kami, Constance, Barbara A. Lewis, and Sally Jones Livingston. "Primary Arithmetic: Children Inventing Their Own Procedures." *Arithmetic Teacher* 41 (December 1993): 200-03.

Kamii, Constance, and Ann Dominick. "The Harmful Effects of Algorithms in Grades 1–4." In *The Teaching and Learning of Algorithms in School Mathematics,* 1998 Yearbook of the National Council of Teachers of Mathematics, edited by Lorna J. Morrow, pp. 130–40. Reston, Va.: National Council of Teachers of Mathematics, 1998.

Kouba, Vicky L., Thomas P. Carpenter, and Jane O. Swafford. "Number and Operations." In *Results from the Fourth Mathematics Assessment of the National Assessment of Educational Progress,* edited by Mary Montgomery Lindquist, pp. 64–93. Reston, Va.: National Council of Teachers of Mathematics, 1989.

Kouba, Vicki L., Judith S. Zawojewski, and Marilyn E. Strutchens. "What Do Students Know about Numbers and Operations?" In *Results from the Sixth Mathematics Assessment of the National Assessment of Educational Progress,* edited by Patricia Ann Kenney and Edward A. Silver, pp. 87–140. Reston, Va.: National Council of Teachers of Mathematics, 1997.

Lampert, Magdalene. "Arithmetic as Problem Solving." *Arithmetic Teacher* 36 (March 1989): 34–36.

Lampert, Magdalene. "Teaching Multiplication." *Journal of Mathematical Behavior* 5, No. 3 (December 1986): 241–80

Leutzinger, Larry P. "Developing Thinking Strategies for Addition Facts." *Teaching Children Mathematics* 6 (September 1999): 14-18.

Moyer, Patricia Seray. "A Remainder of One: Exploring Partitive Division." *Teaching Children Mathematics* 6 (April 2000): 517-521.

Nagel, Nancy G., and Cynthia Carol Swingen. "Students' Explanations of Place Value in Addition and Subtraction." *Teaching Children Mathematics* 5 (November 1998): 164-170.

Reys, Barbara J., and Robert E. Reys. "Computation in the Elementary Curriculum: Shifting the Emphasis." *Teaching Children Mathematics* 5 (December 1998): 236-241.

Russell, Susan Jo. "Developing Computational Fluency with Whole Numbers." *Teaching Children Mathematics* 7 (November 2000): 154-158.

Schultz, James E., and Michael S. Waters. "Why Representations?" *Mathematics Teaching in the Middle School* 6 (September 2000): 448-453.

Schultz, James. "Area Models—Spanning the Mathematics of Grades 3-9." *Arithmetic Teacher* 39 (October 1991): 42-46.

Steffe, Leslie P. "Children's Multiplying Schemes." In *The Development of Multiplicative Reasoning in the Learning of Mathematics,* edited by Guershon Harel and Jere Confrey, pp. 3–39. Albany, N.Y.: State University of New York Press, 1994.

Sundar, Viji K. "Thou Shalt Not Divide by Zero." *Arithmetic Teacher* 37 (March 1990): 50-51.

Thomas, Cynthia S. "100 Activities for the 100th Day." *Teaching Children Mathematics* 6 (January 2000): 276-280.

Thornton, Carol A. "Strategies for the Basic Facts." In *Mathematics for the Young Child,* edited by Joseph N. Payne, pp. 133–51. Reston, Va.: National Council of Teachers of Mathematics, 1990.

Trafton, Paul R., and Christina L. Hartman. "Developing Number Sense and Computational Strategies in Problem-Centered Classrooms." *Teaching Children Mathematics* 4 (December 1997): 230–33.

Vance, James H. "Number Operations from an Algebraic Perspective." *Teaching Children Mathematics* 4 (January 1998): 282-85.

Wearne, Diana, and James Hiebert. "Place Value and Addition and Subtraction." *Arithmetic Teacher* 41 (January 1994): 272-74. *Understanding the procedures used by students.*

Weinberg, Susan. "Going beyond Ten Black Dots." *Teaching Children Mathematics* 2 (March 1996): 432-35.

Books and Other Literature:

Burns, Marilyn, and Bonnie Tank. *A Collection of Math Lessons from Grades 1 through 3* (1988). The Math Solution Publications (Distributed by Cuisenaire Company of America, Inc.).

Carle, Eric. *The Very Hungry Caterpillar.* New York: Putnam, 1994.

Creative Publications Staff. *Hands On Base Ten Blocks* (1986). Creative Publications.

Hope, J., L. Leutzinger, B. Reys, and R. Reys. *Mental Math in the Primary Grades* (1986) Dale Seymour Publications.

De Paola, Tomie. *The Popcorn Book*. New York: Holiday House, 1978.

Friedman, Aileen. *The King's Commissioners*. New York: Scholastic, 1994.

Lankford, Mary D. *Dominoes Around the World*. New York: Morrow Junior Books, 1998.

Madell, Robert. *Picturing Multiplication and Division from Models to Symbols* (1979) Creative Publications, Inc.

Madell, Robert, and Elizabeth Larkin Stahl. *Picturing Addition from Models to Symbols* (1977). Creative Publications, Inc.

Madell, Robert, and Elizabeth Larkin Stahl. *Picturing Numeration from Models to Symbols* (1977). Creative Publications, Inc.

Madell, Robert, and Elizabeth Larkin Stahl. *Picturing Subtraction from Models to Symbols* (1977). Creative Publications, Inc.

McGrath, Barbara Barbieri. *The M & M's Brand® Counting Book*. Watertown, MA: Charlesbridge Publishing, 1994.

McGrath, Barbara Barbieri. *The Cheerios Counting Book*. New York: Scholastic, 1998.

Pittman, Helena C. *A Grain of Rice*. New York: Hastings House, 1986.

Schwartz, David. *How Much Is a Million?* New York: Lothrop, Lee and Shepard, 1985.

Schwartz, David. *If You Made a Million*. New York: Lothrop, Lee and Shepard, 1989.

Schwartz, David. *G is for Googal: A Math Alphabet Book*. Berkeley, CA: Tricycle Press, 1998.

Other Resources:

Burns, Marilyn. *Mathematics: With Manipulatives* (1988). Cuisenaire Company of America, Inc. *Base Ten Blocks Video.*

Burns, Marilyn. *Mathematics: With Manipulatives* (1988). Cuisenaire Company of America, Inc. *Cuisenaire™ Rods Video.*

5 THE SYSTEM OF INTEGERS AND ELEMENTARY NUMBER THEORY

Teaching isn't telling, it's asking. - Miriam Leiva

Students' first experiences with numbers are with the natural or counting numbers and then the whole numbers. However, students often know from their own experiences (below 0° temperatures, losing points in games) that there are numbers that are opposites of whole numbers. Two thousand years ago the Chinese used black and red rods to deal with positive and negative numbers. Students today may use black and red squares to model integers with the black squares representing positive integers and red squares representing negative integers. (See *Modern Mathematics*, Chapter 5.) In this activities manual we will use a hot air balloon model to develop the four operations with integers.

"Number theory offers many rich opportunities for explorations that are interesting, enjoyable, and useful . . . . Challenging but accessible problems from number theory can be easily formulated and explored by students. For example, building rectangular arrays with square tiles can stimulate questions about divisibility and prime, composite, square, even, and odd numbers. . . ." (NCTM *Curriculum and Evaluation Standards, 1989*)

Explorations in number theory reap dividends in problem solving, in the understanding and development of other mathematical concepts, in illustrating the beauty of mathematics, and in understanding the human aspects of the historical development of number. Some problems in number theory which are yet unsolved by mathematicians are simple enough for children to understand. Perhaps that is why this branch of mathematics has intrigued so many people for over 2000 years.

Numbers are special to mathematicians. Some are so special that they have special names: abundant, aliquot part, composite, crowd, deficient, figurate, perfect, practical, prime, square, weird. The first steps in the development of number theory were taken around 550 BC by Pythagoras and his followers in conjunction with the philosophy of their closely-knit brotherhood. Pythagoras founded the famous Pythagorean school which was an academy for the study of philosophy, mathematics, and natural sciences. The Pythagoreans believed that whole numbers were the building blocks of everything. Significance was attached to certain natural numbers. (Eves, *In Mathematical Circles*)

One: the originator of all numbers, the number of REASON
Two: the first even or female number, represented DIVERSITY of OPINION
Three: the first male number, represented HARMONY, composed of unity and diversity
Four: JUSTICE or RETRIBUTION, suggesting the squaring of accounts
Five: the union of first male and female numbers, represented MARRIAGE
Six: the number of CREATION (just as in the Judeo/Christian tradition)

We hope you enjoy reason and harmony and avoid retribution as you complete the activities based on number theory.

Activity 5-1: Adding with the Hot Air Balloon

Overview: Many devices have been used to model the operations with integers. These have included colored chips, movement on the number line, or the operation of a checking account. Several of these models are discussed briefly in *Modern Mathematics;* the colored chip model is discussed in some detail. We are presently going to discuss an alternative model built on the operation of a hot air balloon. This model, developed by Bill Kring, a Presidential Awardee for Excellence in Mathematics Teaching (WA 1992), has proven to be very attractive to young people and it has the additional virtue of allowing us to model multiplication and division as easily and naturally as we model addition and subtraction.

Students' experiences with helium-filled balloons and their seeing these balloons on television give them an intuitive feel for the effects of gas (hot air) and sand on a balloon. Gas expands the balloon (providing lift) which makes it go up; sand adds weight making it go down. In this model, then, gas bags represent positive integers and sand bags represent negative integers. Specifically, each gas bag has enough lift to raise the balloon exactly one unit and each sand bag has enough weight to lower the balloon exactly one unit. It is reasonable to children to take "up" as the positive position and "down" as the negative position. Also, to facilitate description of movement of the balloon, we will take 0 to be the level of the lip of the Grand Canyon. This allows the balloon to float above 0 or sink below 0. Some aspects of this model require you to use your imagination. For example, we assume that there is an unlimited supply of and capacity for gas bags and sand bags, and that there is no problem in the delivery or removal of these bags. Children are quite willing to accept these conditions and to use this model as a powerful learning tool.

Work in pairs on this activity.

Materials: A copy of the two balloon models on Appendix Page 5 and a copy of the pair of vertical number lines on that same page

A. **Introduction to the Hot Air Balloon:** It is important to get an intuitive feel for how the hot air balloon moves. If a gas bag is put on, the balloon rises one unit; if a gas bag is taken off, the balloon falls one unit. Similarly, if a sand bag is put on, the balloon falls one unit; if a sand bag is taken off, the balloon rises one unit. Consider the balloon hovering at the initial position shown below. If three gas bags are put on, it rises to the final position.

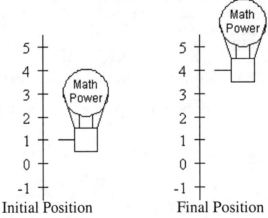

Initial Position   Final Position

Now cut out your copies of the two balloons on Appendix Page 5 and use the vertical number lines on that page. Place the first balloon at the initial position, and the second balloon at the position that results from each of the following operations. After completing each part, compare your results with your partner's.

(a) Initial position: 3
    Put on 2 gas bags

(b) Initial position: 0
    Take off 2 sand bags

(c) Initial position: ¯1
    Put on 2 sand bags

(d) Initial position: 6
    Take off 3 gas bags.

B. (a) If the balloon hovers at an initial position of 3 and we put on 6 gas bags, it will rise to a position of 9 on the number line. Think of two other ways you could arrange for the balloon to rise from 3 to 9. Share your ideas with your partner and then write them below.

(b) Below you will find an initial and final position for a balloon. Think of at least three ways we can put on or take off gas and/or sand bags to accomplish this movement.

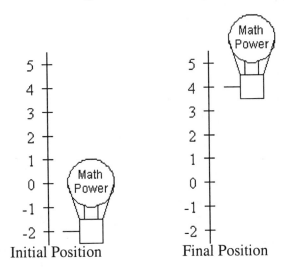

Initial Position   Final Position

TEACHING NOTE:
When using this activity with children, propose a problem like those in Part B on the overhead projector or on the board. Give sufficient time for each student to formulate a solution. Then accept solutions from the class and discuss them. Understanding will increase when one student proposes to take off sand bags to make the balloon go down and others realize that this will make the balloon go up.

C. **Adding with the Balloon Model:**   In the balloon model, addition is associated with the action of "putting on." Thus, an addition problem with integers has three parts.

(1) The first number represents the initial position of the balloon on the number line.
(2) The addition operation following that number indicates that something is being "put on" the balloon. (A subtraction operation indicates that something is "taken off" the balloon.)
(3) The second number determines how many and what type of bags are put on the balloon.

The answer to the addition problem is indicated by the final position of the balloon on the number line.

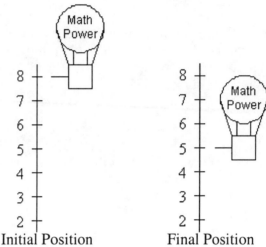

Example: Consider 8 + ¯3. Translate this expression as:

The balloon is at 8, and three sand bags are put on it.

At the right you see the initial and final positions of the balloon. The sum is 5.

Initial Position   Final Position

For each of the following problems, first translate the problem into balloon language (as in the example above) and then use your balloons and your vertical number lines to compute the following sums. Compare answers in each instance with your partner's.

(a)

Problem	Translation	Sum
¯7 + ¯6	Start at ¯7 and put on 6 sand bags.	
3 + ¯5		
¯4 + 3		
¯4 + ¯5		

(b) Consider a problem that is translated: The balloon is at ¯4 and 12 gas bags are put on it. What addition problem does this represent?

(c) In view of your work from the table in (a), what happens every time you put gas bags on the balloon?

(d) What happens every time you put sand bags on the balloon?

TEACHING NOTE:

When this activity is introduced to a class of children, give each child a single balloon, a page with a vertical number line, and a paper clip. Pose a problem on the overhead projector or the board and ask each child to paper clip his or her balloon in the appropriate place on the number line and hold it in the air. When all have had a chance to try a solution, call on children and ask them to translate the problem into balloon language and discuss their solution (perhaps at the board or the overhead projector).

At the end of the activity, solicit general rules from the children about balloon arithmetic. As children formulate rules, it is always good practice to give them ownership of their idea by naming the rule after them. For example, in future discussion in the class, refer to the rule as Kim's rule or Shondea's rule.

TEACHING NOTE:

Review these comments from *Principles and Standards for School Mathematics* that emphasize grade-level specific goals relative to the study of integers.

> Middle-grades students should also work with integers. In lower grades, students may have connected negative integers in appropriate ways to informal knowledge derived from everyday experiences, such as below-zero winter temperatures or lost yards on football plays. In the middle grades, students should extend these initial understandings of integers. Positive and negative integers should be seen as useful for noting relative changes or values. Students can also appreciate the utility of negative integers when they work with equations whose solution requires them, such as $2x + 7 = 1$.

> Negative integers should be introduced at this level through the use of familiar models such as temperature or owing money. The number line is also an appropriate and helpful model, and students should recognize that points to the left of 0 on a horizontal number line can be represented by numbers less than 0.

We're not moving fast enough for our children; we're moving fast enough for us.
- Larry Hatfield

Activity 5-2: Subtracting with the Hot Air Balloon

Overview: If you have carefully completed Activity 5-1, you are ready to complete this activity and understand that each subtraction problem can be rewritten as an addition problem. *Work in pairs on this activity.*

Materials: A copy of the two balloon models on Appendix Page 5 and a copy of the pair of vertical number lines on that same page

A. **Subtracting with the Balloon Model:** In the balloon model subtraction is associated with the action of "taking off." Thus a subtraction problem with integers has three parts.

   (1) The first number represents the initial position of the balloon on the number line.
   (2) The subtraction operation following that number indicates that something is being "taken off" the balloon.
   (3) The second number determines how many and what type of bags are taken off the balloon.

The answer to the subtraction problem is indicated by the final position of the balloon on the number line.

Example: Consider $4 - {}^{-}3$. Translate this expression as:

"The balloon is at 4, and three sand bags are taken off."

At the right you see the initial and final positions of the balloon. The difference is 7.

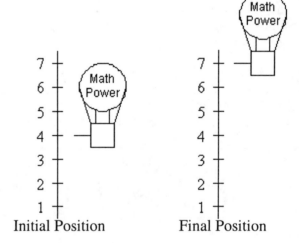

Initial Position   Final Position

For each of the problems in the table on the next page, first translate the problem into balloon language (as in the example above) and then use your balloons and your vertical number lines to compute the sums. Compare answers in each instance with your partner's.

(a)

Problem	Translation	Result
⁻2 – 6	Start at ⁻2 and take off 6 gas bags.	
3 – ⁻5		
⁻4 – 3		
⁻4 – ⁻5		

(b) Consider a problem that is translated: The balloon is at ⁻1 and 7 sand bags are taken off it. What subtraction problem does this represent?

(c) In view of your work from the table in (a), what happens every time you take gas bags off the balloon?

(d) What happens every time you take sand bags off the balloon?

B. **Addition and Subtraction:** Translate each of the following problems into balloon language and then perform the operation with the balloon model. Check each result with your partner's.

(a)

Problem	Translation	Result
⁻3 – 5		
⁻3 + 5		
6 – 2		
6 + ⁻2		

(b) Compare the effect on the balloon of putting on 5 gas bags or taking off 5 sand bags.

(c) State the result from (b) in terms of addition and subtraction.

(d) Compare the effect on the balloon of putting on 2 sand bags or taking off 2 gas bags.

(e) State the result from (d) in terms of addition and subtraction.

Activity 5-3: Multiplication with the Hot Air Balloon

Overview: Remember that one of the basic definitions for multiplication involves repeated addition. In particular, 4 × 5 means *add 4 groups of 5 objects* to get 20 objects altogether. Since 4 means "add four groups," it will be translated "put on four groups" for the balloon model. Since the 5 indicates the number of objects in each group, it means "five gas bags in each group" for the balloon model. Since 20 represents the total number of objects when the groups are pushed together, it indicates where the balloon will finally come to rest after it has started at 0. Proceed and you will become an expert at multiplication on the balloon model.
    Work in pairs on this activity.

Materials: A copy of the two balloon models on Appendix Page 5 and a copy of the pair of vertical number lines on that same page

A. **Multiplying with the Balloon Model:** We will model multiplication on the balloon in the following way:

    (1) The first number represents *how many groups of bags* are involved and *what is being done with them.*
    (2) The second number determines *how many are in each group* and *what kind of bag* is in each group.
    (3) The balloon starts at 0 and the answer is determined by where the balloon comes to rest.

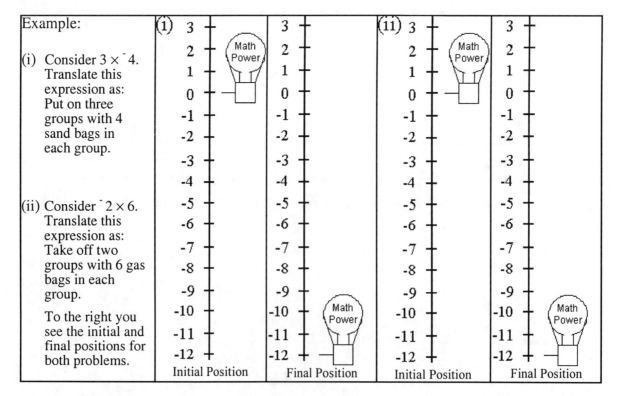

| Example: | (i) Consider 3 × ⁻4. Translate this expression as: Put on three groups with 4 sand bags in each group. (ii) Consider ⁻2 × 6. Translate this expression as: Take off two groups with 6 gas bags in each group. To the right you see the initial and final positions for both problems. |

For each of the following problems, first translate the problem into balloon language (as in the preceding example) and then use your balloons and your vertical number lines to compute the following products. Compare answers in each instance with your partner.

Problem	Translation	Product
⁻2 × ⁻6	Start at 0 and take off 2 groups of 6 sand bags.	
3 × ⁻5		
⁻4 × ⁻3		
⁻4 × ⁻5		

B. Each time you place groups of gas bags on the balloon, the balloon rises from 0. This translates into the following rule about multiplication of integers:

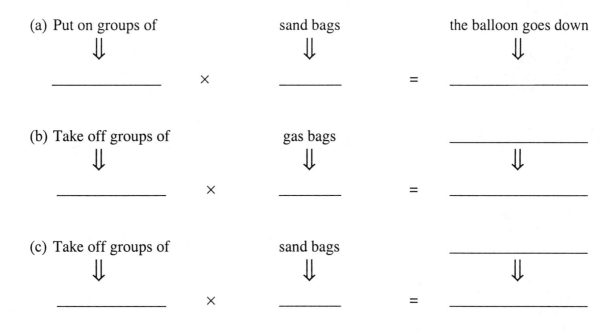

Put on groups of		gas bags		the balloon goes up
⇓		⇓		⇓
positive number	×	*positive number*	=	*positive number*

Complete the description of each of the following balloon operations and describe the associated multiplication pattern for integers that corresponds to it.

(a) Put on groups of   sand bags   the balloon goes down
              ⇓   ⇓   ⇓

    _____   ×   _____   =   _____

(b) Take off groups of   gas bags   _____
              ⇓   ⇓   ⇓

    _____   ×   _____   =   _____

(c) Take off groups of   sand bags   _____
              ⇓   ⇓   ⇓

    _____   ×   _____   =   _____

Activity 5-4: Division with the Hot Air Balloon

Overview: Remember that measurement division is one of the models for division. In the problem 30 ÷ 5, you may ask "How many groups of 5 are in 30?" Alternatively, you may ask: "How many groups of 5 must be added to get 30 objects?" In using the balloon model, we will ask, "If the balloon begins at 0, how does it get to 30 using groups of 5 gas bags?" The answer involves two parts: *what to do with those groups of bags* and *how many times to do it.* To get from 0 to 30, the balloon must go up. Thus, you will have to **put on** *6 groups*: Your answer is + 6. Consider the problem ‾24 ÷ 3. In using the balloon model we will ask, "If the balloon begins at 0, how does it get to ‾24 using groups of 3 gas bags?" Once again, the answer involves two parts: what to do with those groups of bags and how many times to do it. Since the balloon is to go down, you will have to **take off** *8 groups*: your answer is ‾8.
  Work in pairs on this activity.

Materials: A copy of the two balloon models on Appendix Page 5 and a copy of the pair of vertical number lines on that same page

A. **Dividing with the Balloon Model:** We will model division using the balloon in the following way.

(1) The first number represents where the balloon comes to rest after starting at 0.
(2) The second number determines how many and what kind of bag is in each group used to get there.
(3) The answer is what to do with those groups of bags and how many times to do it.

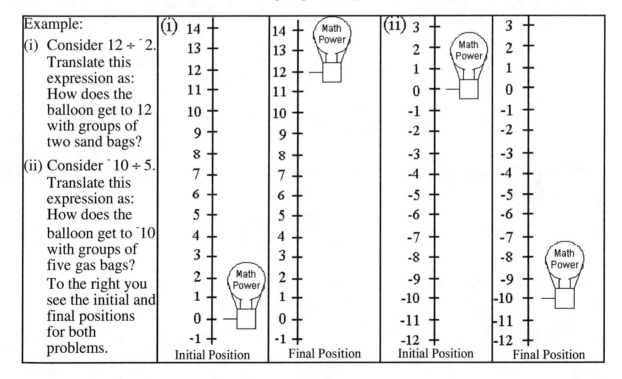

B. For each of the following problems, first translate the problem into balloon language (as in the preceding example) and then use your balloon and your vertical number lines to compute the following quotients. Compare your answers in each instance with your partner's.

Problem	Translation	Quotient
$^-20 \div ^-4$	How do you get to $^-20$ with groups of 4 sand bags?	
$14 \div ^-7$		
$^-9 \div 3$		
$15 \div 5$		

TEACHING NOTE:

In teaching multiplication and division to students using the hot air balloon model, you will, of course, have them work through and model many more examples than in the preceding activities. They need sufficient experience with moving the balloon so that when it comes time to formulate general rules for multiplying integers in Part B of Activity 5-3, they may do so on the basis of their experience. Such results will stay with the children longer and can be recalled more easily. Additionally, you should allow students to make up problems and to explain their reasoning orally and in writing.

Activity 5-5: Factors and Multiples

Overview: This activity reinforces some of the terminology related to number theory and develops understanding using one-inch squares. Note: a 2×5 rectangle made with 10 squares is congruent to a 5×2 rectangle, so the two rectangles are not considered geometrically "different."

Materials: One-inch squares (or square tiles) of assorted colors.

A. **From Rectangles to Factors:**

(a) Using 6 squares, make two different rectangles.
   What are the dimensions of one of the rectangles? _____×_____
   What are the dimensions of the other rectangle? _____×_____

(b) The dimensions of the rectangles made with six squares are the **factors** of 6. List the factors of 6 in ascending order:_____,_____,_____,_____

(c) Find all the factors of 16 by making rectangles.

(d) Find all the factors of 18 by making rectangles. List them.

B. When we write $2 \times 3 = 6$, we can say: "Two and three are factors of six." So, 6 is a **multiple** of 2 and a multiple of 3. There are two related division facts associated with $2 \times 3 = 6$: $6 \div 2 = 3$ and $6 \div 3 = 2$. Therefore, 2 and 3 are also **divisors** of six. Six **is divisible by** 2 and **is divisible by** 3. [Note: In the case of positive numbers, *factor* and *divisor* are synonyms.]

Write statements using the words **factor**, **multiple**, or **divisor**, and the phrase **is divisible by** related to " $7 \times 4 = 28$."

C. (a) Using squares, demonstrate that 3 is a factor of 24.

(b) Using squares, demonstrate that 5 is NOT a factor of 23.

Activity 5-6: Even and Odd

Overview: The counting or natural numbers can be represented geometrically with patterns of squares involving one or two rows.
The activity should be completed in groups of 2.

Materials: One-inch squares and transparent tape (or square tiles to model and grid paper to record and cut out)

A. (a) The first five natural (or counting) numbers are represented below with a geometric pattern. Continue the pattern for 6, 7, 10, and 13. Make these patterns with paper squares and tape (or draw them on grid paper and cut out).

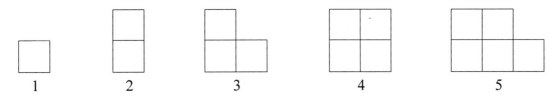

   (b) Describe, to a partner, how the models for 1, 3, 5, 7, and 13 differ from the models for 2, 4, 6, and 10. Those numbers with the characteristic you described for 2, 4, 6, and 10 are called **even** numbers. The others are called **odd** numbers.

   Using your characteristic, describe why 8 is an even number. _____

B. **Modeling Even Numbers:** The first three even numbers are modeled below.

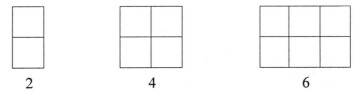

   (a) Draw the figure for the sixth even number.

   (b) What is the sixth even number? _____

   (c) Describe, in words, how the 120th even number would look.

   (d) What is the 120th even number? _____

   (e) Describe, in words, how the nth even number would look.

   (f) What mathematical expression represents the nth even number? ___

C. **Modeling Odd Numbers:** The first three odd numbers are modeled below.

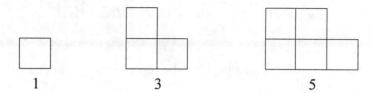

1   3   5

(a) Draw the figures for the sixth and seventh odd numbers.

(b) What is the sixth odd number? _____

(c) What is the seventh odd number? _____

(d) Describe, in words, how the 17th odd number would look.

(e) What is the 17th odd number? _____

(f) Describe how the 40th, 100th, and n^{th} odd numbers would look.

(g) What is the 40th odd number? _____

(h) What is the 100th odd number? _____

(i) What mathematical expression represents the nth odd number? ____

D. Determine whether each of the following is even or odd. Explain your reasoning and justify by drawing pictures to illustrate your explanation.

(a) The sum of two even numbers

(b) The sum of two odd numbers

(c) The sum of an even number and an odd number

(d) The sum of three odd numbers

(e) The sum of any three consecutive natural numbers

Activity 5-7: Figurate Numbers

Overview: Some types of numbers have special geometric patterns: squares, cubes, and triangular numbers are some of these. This activity helps students make connections between special natural numbers and geometry.

Materials: Colored one-inch squares and square grid paper (Appendix Page 6) for recording; wooden or multilink cubes; isometric dot paper (Appendix Page 7); crayons or colored markers

A. **Squares (Square Numbers)**

(a) Put one colored square in your work space. It is a representation for the smallest square that can be made with one-inch squares. It is a 1×1 square and its area is 1 square unit. On your square grid paper, color in one square on the left of the page about half way down.

Using a second color, add squares to the one you already have in your work space to form the next larger square. What are its dimensions? _____
What is its area? _____ How many additional squares did you use? _____ Color these added squares on the grid sheet.

Using a third color, add squares to the one you now have to form the next larger square. What are its dimensions? _____ What is its area? _____ How many additional squares did you use? _____ Color these added squares on the grid sheet.

Continue this pattern four more times.

The numbers (1, 4, 9, 16, 25, 36, . . .) representing the areas of the squares you have formed are called **square numbers** because, geometrically, a figure we call a square is formed.

(b) Your grid sheet can be used to write square numbers in terms of the number of different colored squares that are used to form the squares. For example, the 2×2 square is formed by one square of one color and three squares of another color. The 3×3 square is formed by one square of one color, three squares of a second color, and five squares of a third color. We can write this numerically as indicated in the chart below. Continue the pattern and complete the chart.

Dimensions of Square	Sum of Squares of Different Colors	Area of Square
1×1	1	1
2×2	$1 + 3$	4
3×3	$1 + 3 + 5$	9
4×4		
5×5		
6×6		
7×7		

(c) Without adding each term individually, find the square number given by

$$1 + 3 + 5 + 7 + 9 + 11 + 13 + 15 + 17 + 19 =$$

$$1 + 3 + 5 + 7 + 9 + 11 + \ldots + 31 =$$

B. **Cubes:**

(a) Put a cube in your work space. A cube is a rectangular prism or box with three equal dimensions. It is a $1 \times 1 \times 1$ cube, and its volume is 1 cubic unit.

Add cubes to the one you already have in your work space to form the next larger cube. What are its dimensions? _____ What is its volume?

Add cubes to the one you now have to form the next larger cube. What are its dimensions? _____ What is its volume?

(b) If you were to add more cubes to form the next larger cube, what would the dimensions be? What would the volume be?

(c) How can you find the volume of a cube if one of the dimensions is 5? 6?

(d) The numbers (1, 8, 27, 64, . . .) representing the volumes of the cubes you have formed are called **cubes** because, geometrically, a three-dimensional figure we call a cube is formed. Find three numbers that are both squares and cubes.

C. Triangular Numbers:

(a) Using isometric dot paper, color a single colored dot.

Now, with a different color, color the number of dots necessary to build the next larger triangle. How many dots did you add? What are the total number of dots in the triangle?

With yet another color, color the number of dots necessary to build the next larger triangle. How many dots did you add? What are the total number of dots in the triangle?

Using four additional colors, repeat this operation four times.

The numbers (1, 3, 6, 10, 15, 21, . . .) representing the number of dots in the triangles that you have formed are called **triangular   numbers.**

(b) The patterns on your isometric dot paper can be used to write triangular numbers in terms of the numbers of different colored dots that are used to form the triangles. For example, the second triangle is formed by one dot of one color and two dots of another color. The third triangle is formed by one dot of one color, two dots of a second color, and three dots of a third color. We can write this numerically as indicated in the chart below. Continue the pattern and complete the chart.

Triangular Number	Sum of Dots of Different Colors	Sum of all dots
1	1	1
2	1 + 2	3
3	1 + 2 + 3	6
4		
5		
6		
7		

(c) Write the sum that would represent the tenth triangular number.

(d) Write the sum that would represent the nth triangular number.

TEACHING NOTE:
In addition to figurate numbers, other special numbers, often associated with the Pythagoreans, have delighted many students of number theory: abundant, deficient, and perfect numbers. For example, a number like 15 is called a *deficient number* because the sum of its proper factors is less than the number itself (1 + 3 + 5 = 9 and 9 < 15). A number like 20 is called an *abundant number* because its proper factors have a sum greater than the number itself (1 + 2 + 4 + 5 + 10 = 22 and 22 > 20). A number such as 6 is called a *perfect number* since the sum of the proper factors is equal to the number itself (1 + 2 + 3 = 6). Note the square number 16 is deficient since 1 + 2 + 4 + 8 < 16 while the square number 36 is abundant since 1 + 2 + 3 + 4 + 6 + 9 + 12 + 18 > 36. Thus, we call 16 a deficient square number and 36 an abundant square number. In fact, there are NO 'perfect' square numbers because no square number is also a perfect number. For this reason, we discourage our students from using the terminology 'perfect squares' when discussing square numbers.

Activity 5-8: Prime and Composite Numbers

Overview: Prime numbers are used in the composition and decomposition of many whole numbers. This activity generates definitions of prime and composite numbers by considering the factors of the natural numbers (as opposed to giving a definition for memorization).

Materials: One-inch squares and paper for recording

A. Using squares, form all the different rectangles for each of the natural numbers 1-15. (Remember that squares are special rectangles). Sketch the rectangles. Complete the chart below. All possible rectangles with 6 and 7 squares are shown and information is recorded in the chart.

$$2 \times 3 \qquad\qquad 1 \times 6 \qquad\qquad\qquad 1 \times 7$$

Number	Factors	Number of Distinct Factors
1		
2		
3		
4		
5		
6	1, 2, 3, 6	4
7	1, 7	2
8		
9		
10		
11		
12		
13		
14		
15		

B. Numbers, except for 1, that form exactly one rectangle are called **prime   numbers**. Numbers which form more than one rectangle are called **composite   numbers**. The number 1, in a class by itself, is neither prime nor composite.

   (a) Write an explanation of how you can describe a prime number or a composite number in terms of the *number of factors* of the number.

   (b) How does the number 1 differ from prime or composite numbers?

   (c) Describe square numbers in terms of number of factors.

   (d) Find a number that has exactly five factors.

   (e) 13 is a prime. Reverse the digits: 31. What kind of number is 31? Note that when the digits of one of these prime numbers is reversed, the resulting number is a different prime number. Find three other prime numbers (greater than 11) each of which yields a different prime number when its digits are reversed.

Activity 5-9: Greatest Common Factor and Least Common Multiple

Overview: Greatest common factor (GCF) and least common multiple (LCM) are important concepts that arise in mathematics. Students need to be able to identify common factors and common multiples.

Materials: Cuisenaire® Rods for Part C (or paper strips cut into lengths, 3 cm, 4 cm, 5 cm, 6 cm and 8 cm)

A. **Greatest Common Factor:** The concept of the greatest common factor of two numbers is introduced by listing all factors of each number, identifying those that are common to each list, and choosing the greatest (largest) of the common factors.

Example: Find the greatest common factor of 12 and 16, written GCF(12, 16):

Factors of 12: 1, 2, 3, 4, 6, 12

Factors of 16: 1, 2, 4, 8, 16

Note that 1, 2, and 4 are the common factors and 4 is the greatest common factor. We write: GCF(12, 16) = 4

Following the example above, determine the greatest common factor of the following pairs of numbers.

(a) GCF(20, 35) =

Factors of 20:

Factors of 35:

(b) GCF(24, 36) =

Factors of 24:

Factors of 36:

(c) GCF(15, 16) =

Factors of 15:

Factors of 16:

B. **Least Common Multiple:** The concept of the least common multiple of two numbers is introduced by listing all the natural number multiples of each number, identifying those that are common to each list, and choosing the least (smallest) of the common multiples. It will be the first occurrence of a common multiple.

Example: Find the least common multiple of 4 and 6, written LCM(4, 6):

Multiples of 4: $4 \times 1, 4 \times 2, 4 \times 3, 4 \times 4, \ldots = $ 4, 8, 12, 16, 20, 24, . . .

Multiples of 6: $6 \times 1, 6 \times 2, 6 \times 3, 6 \times 4, \ldots = $ 6, 12, 18, 24, 30, . . .

Note that 12 and 24 are the common multiples in this list and 12 is the least common multiple. We write: LCM(4, 6) = 12.

Using the concept of least common multiple, determine

(a) LCM(6, 8) =
   Multiples of 6:
   Multiples of 8:

(b) LCM(12, 16) =
   Multiples of 12:
   Multiples of 16:

TEACHING NOTE:
In the elementary curriculum, the set of whole numbers is the domain in which students usually work as they develop concepts of and operations on numbers. The multiples of 5, in this domain, would be 5×0, 5×1, 5×2, 5×3, 5×4, ... or 0, 5, 10, 15, 20, ... Thus students observe that 0 is a multiple of 5. (Recall that even though 0 is a factor in $5 \times 0 = 0$, it is NOT a divisor.) So, in the domain of whole numbers, 0 would be a multiple of every other whole number. Trivially, 0 would be the least common multiple. Therefore, when considering least common multiples, we are looking for positive multiples and restrict our domain to the natural numbers.

C. Take groups of 3-rods (light green) and 5-rods (yellow). (You may also cut centimeter strips of lengths 3 cm and 5 cm.) Line the 3-rods up end-to-end alongside a line of 5-rods. The distances at which the ends of the 3-rods and 5-rods match are common multiples of 3 and 5. The shortest distance at which the two match is the least common multiple of 3 and 5.

Using Cuisenaire® Rods or centimeter strips, find the least common multiple of each of the following:

(a) LCM(4, 7)

(b) LCM(5, 6)

(c) LCM(6, 8)

D. Complete the following chart. What do you notice about the numbers in the last two columns?

A	B	GCF(A, B)	LCM(A, B)	GCF(A, B) × LCM(A, B)	A × B
3	4				
5	10				
6	8				
12	16				
20	30				

TEACHING NOTE:

There is a cure for "cancel" fever. It is critical that we require correct mathematical reasoning and not merely accept correct answers. A student may take $\dfrac{16}{64}$ and "cancel" 6's to get $\dfrac{1\!\!/6}{6\!\!/4} = \dfrac{1}{4}$: a correct answer but incorrect mathematics. Or, one may "cancel" 9's to write $\dfrac{1\!\!/9}{9\!\!/5}$ as $\dfrac{1}{5}$. Again, this is a correct answer but incorrect mathematics. Later, in some algebra setting, a student may "cancel" to write $\dfrac{x^2 - 25}{x + 5} = \dfrac{\overset{x}{x\!\!/^2} - \overset{5}{2\!\!/5}}{x\!\!/ + 5\!\!/} = x - 5$. Correct answer; incorrect mathematics. Using a "sneaky name for one" strategy can help students understand why this "cancel" thing came about. For example,

$$\frac{15}{20} = \frac{3 \times 5}{4 \times 5} = \frac{3 \times \cancel{5}}{4 \times \cancel{5}} = \frac{3}{4} \;\left(\text{since } \frac{3}{4} \times 1 = \frac{3}{4}\right).$$

$$\frac{x^2 - 9}{x^2 - x - 6} = \frac{(x + 3)\cancel{(x - 3)}}{(x + 2)\cancel{(x - 3)}} = \frac{x + 3}{x + 2}$$

Likewise a "sneaky name for one" is used to write $\dfrac{2}{3}$ as an equivalent fraction with a denominator of 12: $\dfrac{2}{3} = \dfrac{\square}{12}$, $\dfrac{2}{3} \times \boxed{1} = \dfrac{\square}{12} \;\Rightarrow\; \dfrac{2}{3} \times \boxed{\dfrac{4}{4}} = \dfrac{\boxed{8}}{12}$.

The correct answer is not the most important thing. The most important thing is correct mathematical reasoning, which, when followed by careful work, will generally lead to the correct answer.

Activity 5-10: Hundred Chart Patterns

Overview: This activity allows you to explore relationships of multiples and factors of numbers 1-100.

Materials: 8 copies of the hundred chart found on this page (preferably on transparencies); crayons or markers (or transparency pens)

Take a hundred chart and a crayon or marker. Since every counting number is a multiple of one, if we colored all multiples of one, we would color everything in our chart. Thus, we will exclude one from consideration in this activity. Write a large numeral 2 on the top of the page. Ring the numeral 2 on the chart and then color all the other squares with multiples of 2—the numbers you would say if counting by twos—2, 4, 6, 8, . . . , 100. With another chart and a different color crayon or marker, write the numeral 3 at the top of the page and ring the 3 on the chart. Color all the other squares with multiples of 3: 6, 9, 12, . . . , 99. Continue with new charts and different colors until you have a sheet for 2's, 3's, 5's, 7's . Try to find all the numbers that are not colored on any of these four charts. (Along with the ringed numbers and except for 1, these are the primes that are less than 101.) Take a hundred chart and color in 4 and every multiple of 4. Using different colors than before, do the same for 6, 8, and 9. Display the charts for 2 – 9 and study the patterns.

1	2	3	4	5	6	7	8	9	10
11	12	13	14	15	16	17	18	19	20
21	22	23	24	25	26	27	28	29	30
31	32	33	34	35	36	37	38	39	40
41	42	43	44	45	46	47	48	49	50
51	52	53	54	55	56	57	58	59	60
61	62	63	64	65	66	67	68	69	70
71	72	73	74	75	76	77	78	79	80
81	82	83	84	85	86	87	88	89	90
91	92	93	94	95	96	97	98	99	100

(a) Compare the charts for 2 and 4. Complete the statement below:

   If a number is a multiple of 4, then _____ .

(b) What other patterns do you observe?

Activity 5-11: Palindromes and Patterns

Overview: Palindromes are interesting numbers. A palindrome is a number that reads the same forward or backward such as 77, 404, 5335, or 81318. 312 is not a palindrome, but if we reverse the digits (213) and then add, we get a palindrome. Students can practice addition while looking for visual patterns in this 0 - 99 chart.

$$\begin{array}{r} 312 \\ +213 \\ \hline 525 \end{array}$$

We say that 312 is a 1-step palindrome because we can turn it into a palindrome in one step. 48 is a 2-step palindrome since

$$\begin{array}{r} 48 \\ +84 \\ \hline 132 \end{array} \quad \text{and} \quad \begin{array}{r} 132 \\ +231 \\ \hline 363 \end{array}$$

Materials: Palindrome Chart (0-99); crayons/markers of 7 different colors

(a) Explore the numbers 0 - 99. Color all the numbers that are already palindromes one color. Consider single-digit numbers as palindromes. Color the 1-step palindromes another, and so on, up to 6-step palindromes. What patterns do you find?

Choose a color for each:

already a palindrome	1-step palindrome	4-step palindrome
	2-step palindrome	5-step palindrome
	3-step palindrome	6-step palindrome

0	1	2	3	4	5	6	7	8	9
10	11	12	13	14	15	16	17	18	19
20	21	22	23	24	25	26	27	28	29
30	31	32	33	34	35	36	37	38	39
40	41	42	43	44	45	46	47	48	49
50	51	52	53	54	55	56	57	58	59
60	61	62	63	64	65	66	67	68	69
70	71	72	73	74	75	76	77	78	79
80	81	82	83	84	85	86	87	88	89
90	91	92	93	94	95	96	97	98	99

(b) Find larger numbers which are palindromes.

(c) Think of words that are letter palindromes, like "NOON" or "MOM" or "RACE CAR."

TEACHING NOTES: A SUMMARY

Review these comments from *Principles and Standards for School Mathematics*:

> Throughout their study of numbers, students in grades 3–5 should identify classes of numbers and examine their properties. For example, integers that are divisible by 2 are called *even numbers* and numbers that are produced by multiplying a number by itself are called *square numbers*. Students should recognize that different types of numbers have particular characteristics; for example, square numbers have an odd number of factors and prime numbers have only two factors.
>
> Students can also work with whole numbers in their study of number theory. Tasks, such as the following, involving factors, multiples, prime numbers, and divisibility, can afford opportunities for problem solving and reasoning.
>
> - Explain why the sum of the digits of any multiple of 3 is itself divisible by 3.
> - A number of the form *abcabc* always has several prime-number factors. Which prime numbers are always factors of a number of this form? Why?

There is an underlying structure of mathematics which takes its many separate facets and connects them into a useful, interesting, and logical whole. Many students learn mathematics as isolated facts and procedures which often mystify them. Much of their knowledge of number and number operations is from a "memorize and regurgitate" perspective, and students do not begin to comprehend the underlying structure of the number systems they have studied. Younger students think about numbers as "how many" – representations of simple quantities. Numbers like 8 could be modeled as a set with 8 objects. With the introduction to integers, the concept of a number has to be extended. 8 and ⁻8 both represent quantities, but they also represent "direction" when the position of each is located using a number line model. Additionally, these numbers can be tied to temperatures of 8° above or 8° below zero or to a gain of 8 points or a loss of 8 points in a game. It is important for the teacher to allow students to develop operations on integers by using their knowledge of operations on whole numbers and by using a physical model such as the hot air balloon which consistently models all four operations.

Through number theory and development of number sense, students can come to understand and appreciate mathematics as a coherent body of knowledge. Number theory provides a wonderful opportunity for teachers to encourage students to explore numbers and their relationships. When students look at what happens to even and odd numbers or mixed numbers as they just add, subtract, multiply or divide, then they can explore some challenging and interesting ideas. Prime and composite numbers, divisibility rules (and why they "work"), casting out nines, and patterns in Pascal's Triangle are topics to be explored, also. The world of perfect, abundant, deficient, and amicable numbers as well as figurate numbers have brought a sense of wonder to students of all ages and all times. There is a feeling of a "common thread" when students realize they are investigating ideas in number theory that people have explored for thousands of years.

Teachers at all levels should try to emphasize properties of number systems being studied, so that students can see, for example, that the associative and commutative properties apply for integers (and for all real numbers) as well as for natural and whole numbers. It is through learning to reason mathematically (one of the five process standards of the NCTM *Principles and Standards for School Mathematics*), that students are enabled to analyze situations, discover patterns, make conjectures about their discoveries, and verify or disprove their theories. Number theory offers an arena in which students can become empowered mathematically as they begin to understand what DOING mathematics is all about.

RESOURCES

Periodicals:

Cemen, Pamela B. "Adding and Subtracting Integers on the Number Line." *Arithmetic Teacher* 40 (March 1993): 388-389.

Christiansen, Evelyn B. "Pythagorean Triples Served for Supper." *Mathematics Teaching in the Middle School* 3 (September 1997) 60-62.

Hopkins, Martha H. "Number Facts--or Fantasy?" *Arithmetic Teacher* 34 (March 1987): 38-42.

Kelsey, Kenneth, and David King. *The Ultimate Book of Number Puzzles*. New York: Barnes and Noble, 1992.

Kent, Laura Brinker. "Connecting Integers to Meaningful Contexts." *Mathematics Teaching in the Middle School* 6 (September 2000): 62-66.

Norman, F. Alexander. "Figurate Numbers In the Classroom." *Arithmetic Teacher* 38 (March 1991): 42-45.

Sherrill, James M. "Magic Squares and Magic Triangles." *Arithmetic Teacher* 35 (October 1987): 44-47.

Snape, Charles, and Heather Scott. *Puzzles, Mazes and Numbers*. London, England: Cambridge University Press, 1995.

"Triangular Numbers in Problem Solving." *The Mathematics Teacher* 92 (December 1999): 820-824.

Wilson, Patricia S. "Zero: A Special Case." *Mathematics Teaching in the Middle School* 6 (January 2001): 300-303, 308, 309.

Books and Other Literature:

Dickinson, Rebecca. *The 13 Nights of Halloween*. New York: Scholastic, Inc., 1996.

Enzenberger, Hans Magnus. *The Number Devil, A Mathematical Adventure*. New York: Henry Holt and Company, 1998.

Eves, Howard. *In Mathematical Circles* (1969). Prindle, Weber, Schmidt.

Garland, Trudi Hammel. *Fascinating Fibonaccis: Mystery and Magic in Numbers* (1987). Dale Seymour Publications.

Garland, Trudi Hammel et al. *Fibonacci Fun: Fascinating Activities with Intriguing Numbers*. Dale Seymour Publications, 1998.

Garland, Trudi Hammel. *Math and Music: Harmonious Connections*. Dale Seymour Publications, 1995.

Garland, Trudi Hammel. *Mathematical Footprints: Discovering Mathematical Impressions All Around Us*. Wide World Publishing/Tetra, 2000.

Scieszka, Jon, and Lane Smith. *Math Curse*. New York: Viking Press, 1995.

Terban, Marvin. *Too Hot to Hoot: Funny Palindrome Riddles*. New York: Clarion Books, 1985.

6 | *INTRODUCTION TO THE RATIONAL NUMBERS*

Development and growth occurs through an enormously complicated and continuous process of interaction with the environment. Through activity the person discovers understanding by re-inventing what he or she wants to understand. This process takes time. - Jean Piaget

Just as the concept of whole number and the concepts of operations on whole numbers must be developed using concrete models before computation is taught, so it is with fractions and decimals. Even though numerous hours are devoted to computation with fractions prior to leaving the middle grades, students still have poor concepts of fractions and operations on them. Reports from national and international assessments point out that students have difficulty learning concepts of fractions and decimals (Post, 1989; Grossman, 1983). Researchers indicate that students have difficulty solving problems involving fractions when physical or pictorial models are used. This suggests that elementary and middle grade students are not given sufficient experience with fractions before symbolic work is done. Students do not recognize that decimals are a way to represent fractions whose denominators are powers of ten, nor do they recognize that money amounts like $2.45 are decimals. Fractions and decimals must be interwoven in teaching.

The transition from whole numbers to fractions and decimals can be difficult for students. Although they may multiply the numerators and then the denominators, for example, they often do not understand why a similar procedure does not work in adding fractions. Concrete or representational models can help students clarify these anomalies. (*NCTM Curriculum and Evaluation Standards,* 1989)

Since children in grades K - 4 develop concepts slowly, the curriculum should provide time for them to construct accurate and stable fraction and decimal concepts. Even though it is suggested that paper-and-pencil computation with fractions and decimals be delayed until later grades, children should extensively explore equivalent fractions, operations using models of fractions and decimals, and the relationship between fractions and decimals. Further, fraction ideas should be integrated into teaching units involving measurement. (Food and money go a long way in motivating children—and their teachers!)

Activity 6-1: In the Beginning

Overview: Two basic ideas should be emphasized when introducing the concept of fractions:

(1) Start with a **unit region**. To represent fractions, this unit region is partitioned into some number of parts, all of the same size; and some (or maybe all or none) of the parts are singled out. Examples are area or length models. In the early stages of fraction concept development, the equal-sized parts should be congruent.

(2) Start with a **discrete set** of objects. To represent fractions, this set is partitioned into some number of subsets, each with the same number of elements; and some (or maybe all or none) of the subsets are singled out.

In each case, two whole numbers are involved. One of the numbers tells the number of parts into which the region (or set) is separated — the **denominator**. The other number tells how many of those parts are singled out — the **numerator**.

This activity should be completed in pairs.

Materials: Paper squares and scissors; circles, squares, rectangles, and equilateral triangles (easily cut with an Ellison Letter Machine™)

A. (a) Take one each of the circle, square, rectangle, and triangle. Fold into two equal-sized parts. Shade one of the two equal-sized parts. You have shaded one-half of the unit region.

   (b) Compare your half of the circular region to someone else's. How are they alike/different?

   (c) Find more than one way to divide the square region into halves. Compare with a partner.

   (d) Do the same for the rectangular and then the triangular regions. Why do each of these models represent halves?

B. Take a paper square and fold into four equal parts as shown below.

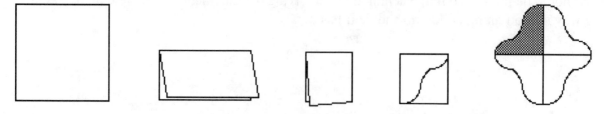

Cut along the non-folded edges, unfold the paper and draw along the creases. This shape is the unit region: one whole. Into how many equal-sized parts is the unit region divided? ____

Shade one of them. Since one of the four equal-sized parts is shaded, we say $\frac{1}{4}$ of the region is shaded. If a unit region is divided into equal-sized parts, we can name and write a fraction to designate the amount which has been shaded.

TEACHING NOTE:
For each region, the teacher should ask the following questions to help students connect the concept of fraction to the name for the fractional part. Expect answers to be given in sentences. Into how many equal-sized parts has the region been divided? (The region has been divided into ___ equal-sized parts.) How many of the equal-sized parts are shaded? (___ of the ___ equal-sized parts are shaded.)

C. Write the fraction for the shaded part of each region.

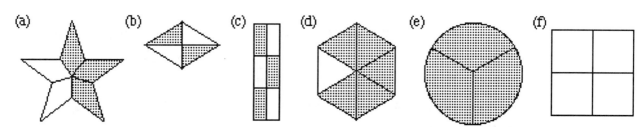

D. In each of the following sets of figures, ring the correct model for the given fraction.

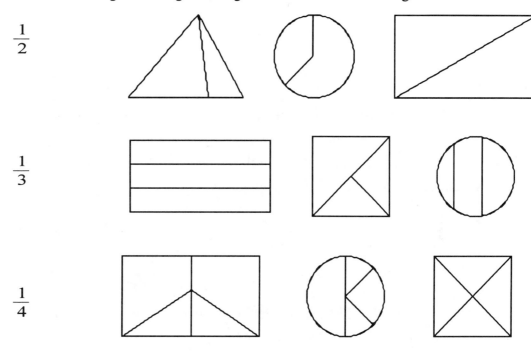

TEACHING NOTE:
When modeling fractions for younger children, divide unit regions into parts which have the same size and shape (congruent parts). In the upper elementary grades, as children become more proficient with the concept of area, one can divide unit regions into parts that have equal area, but perhaps not the same shape. See Activities 6-9 and 6-17.

Activity 6-2: Let's Be Discrete

Overview: This activity develops the concept of fraction with the discrete set model. This model is "natural" to children as they "divvy-up, fair shares."

Materials: "Cut outs" of teddy bears, apples, flowers, pigs, sailboats, *etc.* in assorted colors (easily cut with an Ellison Letter Machine™) or teddy bear counters, two color counters, or colored blocks

A. From a group of counters, select two blue ⬤ ⬤ and one yellow ◯ and put them in your work space. How many counters do you have? _____ How many of the counters are blue? _____ yellow? _____ Complete this sentence: _____ of the 3 counters are blue. Mathematicians would say: "**Two-thirds** of the counters are blue" (since **two** of the **three** counters are blue).

B. From a group of counters, select two blue ⬤ ⬤ and three red ☺ ☺ ☺ and put them in your work space. How many counters do you have? _____ How many of the counters are blue? _____ red? _____ Complete this sentence: _____ of the 5 counters are blue. Mathematicians would say: "**Two-fifths** of the counters are blue" (since **two** of the **five** counters are blue). What fraction of the counters is red?

C. From a group of counters, select two blue ⬤ ⬤ and two yellow ◯◯ and put them in your work space. How many counters do you have? _____ How many of the counters are blue? _____ yellow? _____ Complete this sentence: _____ of the 4 counters are blue. Mathematicians would say: "**Two-fourths** of the counters are blue" (since **two** of the **four** counters are blue). What fraction of the counters is yellow?

Using the same counters, put the blue ones into one group and the yellow ones into another group [you have divided the set into equal-sized parts or equivalent subsets].

Into how many equal-sized parts (or sets) have your counters been divided? _____ How many of the parts are blue? _____ Since **one** of the **two** equal-sized sets is blue, mathematicians would say: "**One-half** of the counters is blue."

D. Take four red counters ⊕ ⊕ ⊕ ⊕ and put in one group.

What fraction of the counters is red? _____ green? _____

Separate the four red counters into two equal-sized groups.

What fraction of the counters is red? _____ green? _____

E. Using 6 counters, show $\frac{1}{2}$ red and $\frac{1}{2}$ green.

F. Using 12 counters, show $\frac{1}{3}$ yellow and $\frac{2}{3}$ blue.

G. A set of counters is red and yellow. $\frac{2}{5}$ of the counters are red and $\frac{3}{5}$ of the counters are yellow. There are six red counters. Model this fraction situation. How many counters are yellow? _____

H. 'Cut outs' of tiny blue teddy bears and pink hearts can be affixed to half-sheets of white
    construction paper to make discrete set fractions cards.

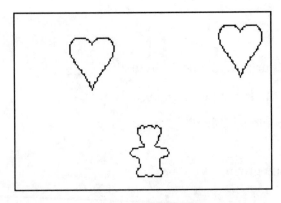

How many shapes are there? _____

How many shapes are hearts? _____

What fraction is hearts? _____

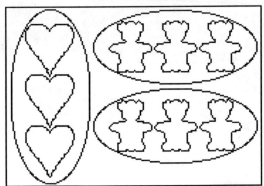

Into how many equal-sized parts are the
shapes divided? _____

____ of the 3 parts is hearts.

What fraction of the shapes is hearts? _____

I. Each discrete set below has twelve shapes: 6 bears and 6 hearts.

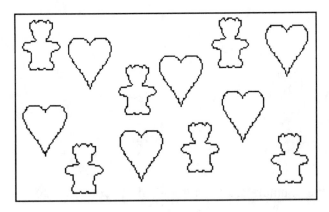

How many shapes are there? _____

____ of the 12 shapes are hearts.

What fraction of the shapes is hearts? ___

What fraction of the shapes is bears? ____

Into how many equal-sized parts are the shapes divided? _____

____ of the ____ parts is hearts.

____ of the ____ parts is bears.

What fraction is hearts? ____ bears? ____

Into how many equal-sized parts are the shapes divided? _____

____ of the ____ parts is hearts.

____ of the ____ parts is bears.

What fraction is hearts? ____ bears? ____

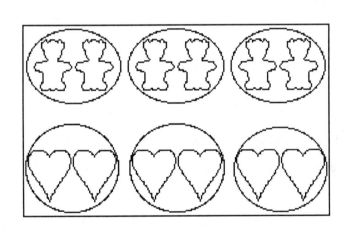

Into how many equal-sized parts are the shapes divided? _____

____ of the ____ parts is hearts.

____ of the ____ parts is bears.

What fraction is hearts? ____ bears? ____

TEACHING NOTE:

In the discrete set fraction cards of Part I, the number of hearts (6) and bears (6) does not change but each fraction is determined by the grouping. This fact demonstrates the equivalence of $\frac{6}{12}$, $\frac{1}{2}$, $\frac{2}{4}$, and $\frac{3}{6}$.

Activity 6-3: Sneaky Names For One (and for other fractions)

Overview: The Circle Fraction Model (or Square Fraction Model) is a useful tool in helping develop the conceptual basis of work with fractions. The Circle Fraction Model consists of 9 circles each cut into different numbers of congruent pieces. Each circle should be a different color so that younger children can call them by their color names. We will use these colors to refer to the various fraction parts.

Unit region	- blue	Sixths	- red
Halves	- green	Eighths	- yellow
Thirds	- pink	Tenths	- light blue
Fourths	- orange	Twelfths	- butterscotch
Fifths	- purple		

The Square Fraction Model is similar to the Circle Fraction Model except it consists of nine squares divided into congruent parts.

Work with a partner.

Materials: Circle Fraction Model or Square Fraction Model (Either of these models can be easily cut out using the appropriate dies with the Ellison Letter Machine™; alternatively, there are black- line masters for the two models on Appendix Pages 8 and 9). If commercial models such as Fraction Factory are available, they may be used instead.

A. Take the blue circular region. Cover with green pieces. $\bigcirc \Rightarrow \bigoplus$

How many green pieces does it take to cover the blue unit?_____ What is the fraction

name for each green piece? _____ How many pink pieces does it take to cover the unit

region?_____ What is the fraction name for each pink piece? _____

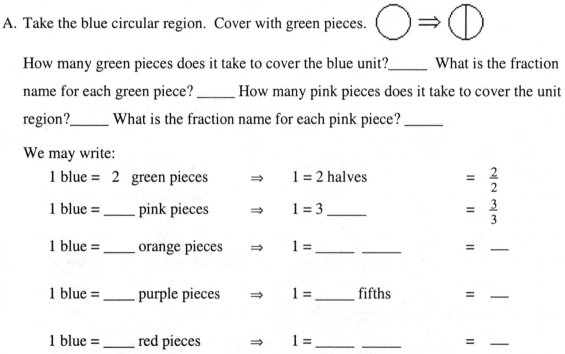

We may write:

1 blue = 2   green pieces $\Rightarrow$ 1 = 2 halves $= \frac{2}{2}$

1 blue = ____ pink pieces $\Rightarrow$ 1 = 3 _____ $= \frac{3}{3}$

1 blue = ____ orange pieces $\Rightarrow$ 1 = _____ _____ $= \underline{}$

1 blue = ____ purple pieces $\Rightarrow$ 1 = _____ fifths $= \underline{}$

1 blue = ____ red pieces $\Rightarrow$ 1 = _____ _____ $= \underline{}$

1 blue = ____ yellow pieces   ⇒   1 = ____ ____   = __

1 blue = ____ light blues   ⇒   1 = ____ ____   = __

1 blue = ____ butterscotches   ⇒   1 = ____ ____   = __

We can say that $\frac{2}{2}$ or $\frac{3}{3}$ or $\frac{6}{6}$ or $\frac{10}{10}$ are "sneaky names for one." When will a fraction be a "sneaky name for one"?

B. Put a green $\frac{1}{2}$ piece in your work space. Cover the green $\frac{1}{2}$ piece with other pieces of one color. Find all the names for $\frac{1}{2}$ using this model. State a generalization for deriving fractions equivalent to $\frac{1}{2}$. Repeat the activity for thirds, fourths, fifths, sixths.

C. **Fraction Building - A Game:** Get a partner. Using either the Circle or Square Fraction Model, make a fraction $\left(\text{like } \frac{3}{4}\right)$ and have your partner match it with at least one equivalent fraction. Score one point for each equivalent fraction that is found. Change roles after each turn. The first person to score 10 points wins. [What strategies did you use so that your chances of winning were greater?]

D. (a) Make 6 different circles using more than one color of fraction pieces. Record your solution for each circle as an addition sentence. *Note: In building your circle, you may not have gaps or overlaps; check your solution by finding a common denominator and adding fractions to make sure you have a fraction equivalent to 1.*
    Example:

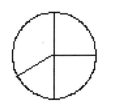 $\frac{1}{4}+\frac{1}{4}+\frac{1}{6}+\frac{1}{3} = 1$ or $\frac{3}{12}+\frac{3}{12}+\frac{2}{12}+\frac{4}{12} = \frac{12}{12} = 1$

Have a partner build your circles by reading your addition sentences.

(b) Write an addition sentence for the diagram below: _____

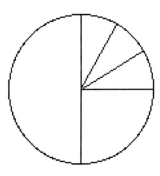

Activity 6-4: Transparent Fractions

Overview: Students are often asked to look at pictorial representations of fractions to identify those which are equivalent. When the pictures are drawn on paper side by side, some students may have difficulty "seeing" that shaded regions are the same. Using transparent grids, students can overlay one on another and readily convince themselves that the two regions are congruent and, thus, the fractions representing those regions are equivalent. Transparent grids can also be overlaid to compare fractions and to multiply fractions.

Materials: Colored pens for transparencies, transparencies made from blackline masters for fraction grids on Appendix Page 10 or Appendix Page 12 (Cut out each square region.)

A. **Equivalent Fractions:**

(a) Take the square region divided into two congruent parts and color

one of the parts red with a transparency marker. Write the fraction

of the square which is red. _____ Overlay the square divided into

four parts. What fraction of the region is red? _____ Has the

amount of colored area changed? _____ We say that $\frac{1}{2}$ and $\frac{2}{4}$ are

different names for the same colored area and we call them

equivalent fractions. Remove the square representing fourths and

overlay the square divided into six parts. What fraction represents

the colored area? _____ Repeat with the squares divided into

eight, ten, and then twelve parts. What fractions represent the

colored area? _____ Do the same for the 10×10 grid.

What fraction represents the colored area now? _____ Write this

fraction as a decimal numeral and then as a percent. _____

(b) Write the names of all the fractions that you found to be equivalent to $\frac{1}{2}$.

(c) Take the square region divided into eight parts and color $\frac{3}{8}$. Can you find a transparent

grid to overlay and determine a fraction equivalent to $\frac{3}{8}$?   Why or why not?

B. **Comparing fractions:**

(a) Make a $\frac{1}{4}$ fraction grid by coloring one of the four parts orange. Make a $\frac{1}{6}$ fraction grid by coloring one of the six parts red. Compare the two shaded regions. Which colored area is larger?

We say that $\frac{1}{4} > \frac{1}{6}$ since the shaded area for $\frac{1}{4}$ is larger than the shaded area for $\frac{1}{6}$.

(b) Using different colors, make a fraction grid for $\frac{3}{4}$ and one for $\frac{2}{3}$. Which fraction is greater? _____ How much greater? _____ Justify using transparent grids. Which fraction is greater: $\frac{2}{3}$ or $\frac{7}{10}$?

C. (a) Make a fraction grid for $\frac{1}{4}$ as shown to the right.

What would $\frac{1}{2}$ of this $\frac{1}{4}$ strip be? _____

Make a $\frac{1}{2}$ fraction grid using a different color and turn the $\frac{1}{2}$ grid at right angles to the $\frac{1}{4}$ grid and overlay it. Into how many parts is this new fraction grid divided? _____ Note that the double-colored region is the same as $\frac{1}{2}$ of the $\frac{1}{4}$ strip we have already considered. We can see that $\frac{1}{2}$ of $\frac{1}{4}$ is $\frac{1}{8}$ and we write: $\frac{1}{2} \times \frac{1}{4}$ is $\frac{1}{8}$.

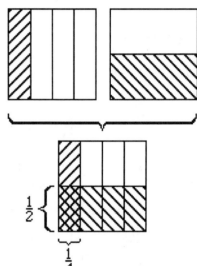

(b) Using different colors, make a $\frac{1}{2}$ fraction grid and a $\frac{3}{8}$ fraction grid. Find $\frac{1}{2}$ of $\frac{3}{8}$ or $\frac{1}{2} \times \frac{3}{8}$ as was done above. Record your answer in the table below. Using fraction grids, model each of the remaining products and record answers in the table.

$\frac{1}{2} \times \frac{1}{4}$	$\frac{1}{8}$
$\frac{1}{2} \times \frac{3}{8}$	
$\frac{1}{8} \times \frac{2}{3}$	
$\frac{1}{6} \times \frac{3}{4}$	
$\frac{3}{10} \times \frac{5}{8}$	
$\frac{3}{4} \times \frac{2}{3}$	

(c) Look for a pattern and state a rule for multiplying two fractions.

(d) Write the multiplication problem and
answer for each picture to the right.

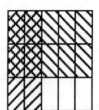
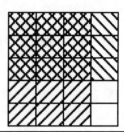

TEACHING NOTE:

Reduce $\dfrac{10}{20}$: $\left(\dfrac{10}{20}\right)$. Reduce it again: $\frac{10}{20}$. Reduce it again: $\frac{10}{20}$.

Perry, one of my mathematically talented fifth graders, had no difficulty taking a problem such as $\frac{5}{7}$ ◯ $\frac{11}{15}$ and determining whether <, >, or = would make the sentence true. But, Perry was missing problems like $\frac{15}{20}$ ◯ $\frac{3}{4}$; when he compared the two fractions he wrote $\frac{15}{20} > \frac{3}{4}$. I couldn't fathom it, so I asked him to explain a problem to me — write $\frac{6}{18}$ in lowest terms. Perry wrote $\frac{6}{18} = \frac{1}{3}$ and said, "$\frac{6}{18}$ equals $\frac{1}{3}$." I asked him why. "I divided the numerator and denominator both by 6; that's really a sneaky name for one." I asked what he could say about the two fractions. "They are equivalent." I asked which fraction was smaller. Looking a little disgusted with me for asking so many questions, he replied, "$\frac{1}{3}$ is smaller." Again, I asked why. Perry: "It's been *reduced*." After interviewing other students in the class, it turned out that four others shared Perry's reasoning. We are not reducing fractions (reducing generally means to make smaller). Rather, we are using smaller numbers to write an equivalent ratio. I learned something about the language we use (or should not use) by listening to a fifth grader that day. Since then, I do not use "reduce" fractions and discourage my students from saying "reduce" when it comes to writing equivalent fractions. Instead I say: "Write in simplest form." This phrase has meaning with more complicated entities such as $\frac{0.5}{0.6}$, $\frac{\frac{1}{2}}{7}$, or $\frac{x+y}{x^2-y^2}$. Also, I discourage my students from saying "Write in lowest terms," since this implies that final numerators and denominators will be smaller than the original parts of the fraction. Consider the examples above: $\frac{0.5}{0.6}$ and $\frac{\frac{1}{2}}{7}$. We change each of them to a common fraction in simplest form using a process that involves multiplying by a form of 1: $\frac{0.5}{0.6} = \frac{0.5}{0.6} \times \boxed{\frac{10}{10}} = \frac{5}{6}$ $\frac{\frac{1}{2}}{7} = \frac{\frac{1}{2}}{7} \times \boxed{\frac{2}{2}} = \frac{1}{14}$.

Notice that the numerator and denominator of $\frac{0.5}{0.6}$ are 10 times as big as they were in $\frac{0.5}{0.6}$, and the numerator and denominator of $\frac{1}{14}$ are 2 times as big as they were in the original fraction $\frac{\frac{1}{2}}{7}$.

Activity 6-5: Ordering Fractions

Overview: A fraction whose numerator is 1 and whose denominator is a nonzero integer is called a **unit fraction**. Students often get confused as to how to tell when one unit fraction is larger or smaller than another (often selecting the one with the larger denominator to be the larger fraction). This activity allows students to see the inverse relationship between the denominator of a fraction and the actual size of each fractional part of a whole.

Work in pairs.

Materials: Circle Fraction Model (Appendix Page 8), Square Fraction Model (Appendix Page 9), fraction strips, or other fraction materials

A. Order one of each fraction piece from largest to smallest.

$$\frac{1}{2} \qquad \frac{1}{3} \qquad \ldots \qquad \ldots \qquad \frac{1}{12}$$

Write the fraction name for each piece in order from largest to smallest. Discuss with a partner how many of each fraction piece is needed to make 1, the unit region. Explain the relationship between the denominator of the fraction and the actual size of each fractional part.

Where would $\frac{1}{9}$ be placed in the sequence?_____

Where would $\frac{1}{11}$ be placed in the sequence?_____

B. Using fraction pieces, model each fraction to compare.

Which fraction is larger: $\frac{3}{5}$ or $\frac{3}{8}$?

Which fraction is larger: $\frac{7}{8}$ or $\frac{7}{10}$?

Which fraction is larger: $\frac{5}{8}$ or $\frac{5}{12}$?

If two fractions have the same numerator but different denominators, how can you tell which is larger?

Activity 6-6: Fractions with Pattern Blocks

Overview: Pattern blocks are geometric shapes that fascinate students of all ages. The set has six shapes: a yellow hexagon, a red trapezoid, an orange square, a green equilateral triangle, a blue rhombus and a tan (skinny) rhombus. The hexagon is regular, the triangle is equilateral, the trapezoid is isosceles. The trapezoid has a longer edge of 2 inches. All the other edges of the shapes measure 1 inch in length. Each block is 1 cm thick.

Materials: Pattern blocks
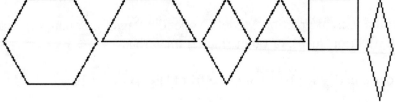

(Pattern blocks can be cut from construction paper with the Ellison Letter Machine™)

A. (a) Let the hexagon represent 1 (the unit region). Cover the hexagon exactly with blocks of the same color.

Can the hexagon be covered with red trapezoids?   How many?

Can the hexagon be covered with blue rhombuses?   How many?

Can the hexagon be covered with green triangles?   How many?

Can the hexagon be covered with orange squares?   How many?

Can the hexagon be covered with tan rhombuses?   How many?

(b) Justify to a partner why the red trapezoid is $\frac{1}{2}$, the blue rhombus is $\frac{1}{3}$, and the green triangle is $\frac{1}{6}$.

TEACHING NOTE:
Recall that area is a measure of covering. Thus, the area of the hexagon is 2 trapezoids, 3 blue rhombuses, or 6 triangles.

(c) Cover the yellow hexagon with blocks of more than one color. How many different ways can it be done? Write a fraction sentence connected to each covering. For example,

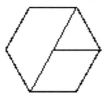

$$1 = \frac{1}{2} + \frac{1}{3} + \frac{1}{6}$$

B. If the square is one-fourth, you can make a whole by putting your four squares together. You may recognize the following five shapes as tetrominoes: all possible arrangements of four squares in a plane matched edge-to-edge. (See Activity 11-9)

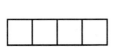

(a) If the green triangle is one-fourth, make a whole. Draw your possible solutions.

(b) If the blue rhombus is one-fourth, make a whole. Draw your possible solutions.

C. **Challenge Problem:** Using pattern blocks, make the shape on the left. Make the same shape beside it using four blocks. By comparing the two shapes, determine how the tan rhombus and the square are related.

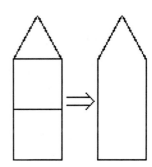

Activity 6-7: Circle Fraction Wheel

Overview: The ability to estimate fractional amounts is helpful in everyday life. Too often, fractions are introduced and taught only at the symbolic level rather than also utilizing the concrete level. Rules and procedures take precedent over concepts and understanding. However, number sense is developed as the arithmetic of fractions is explored with physical or visual models.

Work in pairs on this activity.

Materials: Circle Fraction Model (Appendix Page 8) and two large circular regions of contrasting colors.

A. Take the two circular regions and mark the center of each. Make a cut from the edge to the center and slide one region into the other as shown in the diagram.

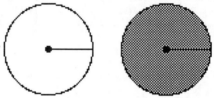

Adjust the circle wheel to estimate $\frac{1}{3}$ and check your estimate using the circle fraction model piece for $\frac{1}{3}$.

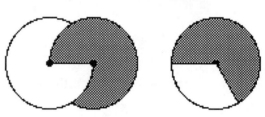

Do the same for $\frac{1}{4}$ and for $\frac{1}{5}$ and check your estimate using the circle fraction model pieces for $\frac{1}{4}$ and $\frac{1}{5}$. (If you turn the circle wheel over for the previous three estimates, you will "see" $\frac{2}{3}$, $\frac{3}{4}$, and $\frac{4}{5}$, respectively.)

B. Discuss with a partner the most efficient way to estimate and check $\frac{11}{12}$ using the circle wheel. What is the most efficient way to estimate and check $\frac{5}{12}$ with the circle wheel?

Activity 6-8: Circle Three

Overview: This game provides a fun way to experience the Circle Fraction Model.
   Play with a partner (or in teams of two).

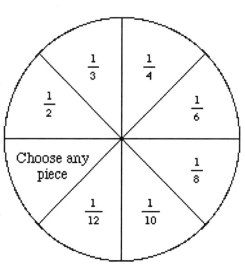

Materials: The Circle Fraction Model (Appendix Page 8), a spinner divided into eighths and marked as shown to the right, the three game circles below for each player or team of players. (Note: With a half-opened paper clip and a pencil, you can make a spinner from the template to the right. See Activity 9-1.) Note that each sector of the spinner represents one-eighth of the area of the circular region, whereas the fraction name in the sector represents which piece of the Circle Fraction Model a player will select to use in playing the game.

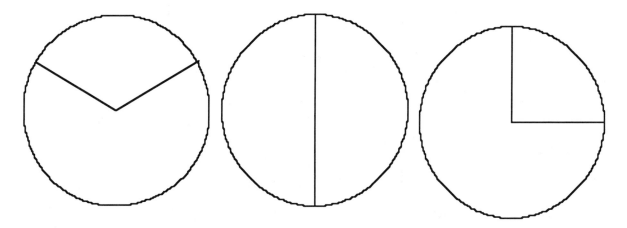

Each player or team of players has to completely fill each of the three circles with fraction pieces from the Circle Fraction Model (the game circles are congruent to the unit circle in the Circle Fraction Model). Fraction pieces cannot cross over the dividing segments in the game circles. A player spins and takes the fraction piece indicated. The fraction piece can be placed anywhere on a game circle as long as it does not cross over a dividing segment and does not overlap another piece already played. Once a piece has been placed, it cannot be moved. If there is not a place for a piece to be played, the turn goes to the other player. The first player to completely fill all three game circles wins. After the game, discuss strategies for determining which piece(s) should go where.

Activity 6-9: Eight Is Enough

Overview: It is important for students to see fractions in a variety of settings. The connections of fractions to geometry and area are enriched by using the geoboard. Additionally, students need to be "creative" in dividing up regions into fractional parts.
Work with a partner.

Materials: Plain paper, a 5 × 5 geoboard (or part of a larger geoboard), and 5 × 5 geoboard dot paper (Appendix Page 11)

A. Individually, draw several squares on your paper. Using straight line segments, divide each square into eighths, with each of the parts being congruent. Compare your results with those of your group. In your class, let each group, in turn, share a solution until the whole class has exhausted the different solutions. Two examples are shown below.

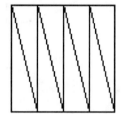   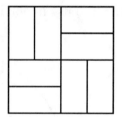

B. Outline a square with area 16 square units on a 5 × 5 geoboard. Using geobands (a.k.a. rubber bands), divide the square into polygons with each representing an area of one-eighth of the larger square and

   (a) the "eighths" are all congruent polygons (see example (i) below).

   (b) the "eighths" need not be congruent polygons (see example (ii) below).

(i)   (ii)

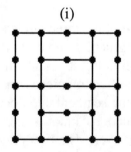   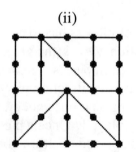

Record your solutions on geoboard dot paper.

Activity 6-10: Strip It!

Overview: Folding a paper strip provides easy access to the concept of equivalent fractions.

Materials: A strip of paper from a register tape - about half a yard will do (One $8\frac{1}{2} \times 14$ sheet cut into fourths lengthwise can be shared by four students.)

A. (a) Consider the length of the strip as a unit. Fold the strip by matching the two ends and label the crease (near the top of the strip) as $\frac{1}{2}$. Refold along the crease and fold in half again. Unfold the strip and label the creases as $\frac{1}{4}$, $\frac{2}{4}$, and $\frac{3}{4}$.

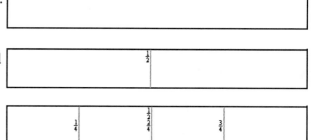

(You'll need to write "$\frac{2}{4}$" beneath the "$\frac{1}{2}$" on the strip.)

Refold and fold in half again. Unfold and mark each crease as $\frac{1}{8}$, $\frac{2}{8}$, $\frac{3}{8}$, $\frac{4}{8}$, $\frac{5}{8}$, $\frac{6}{8}$, and $\frac{7}{8}$.

Notice that some creases (or distances) have different names. For example, $\frac{1}{2}$ and $\frac{2}{4}$ and $\frac{4}{8}$ name the same distance. Which other distances have different names?

(b) Unfold the strip and fold it into thirds, marking the creases as $\frac{1}{3}$ and $\frac{2}{3}$. Re-fold into thirds and then fold in half. How will these creases be marked? _____

Refold and fold in half again. How will these creases be marked _____

(c) Reading from the fraction strip, write all fractions equivalent to

$\frac{1}{2}$:

$\frac{1}{3}$:

$\frac{2}{3}$:

$\frac{3}{4}$:

B. Use the strip to answer these questions.

   (a) Which distance is greater: $\frac{2}{3}$ or $\frac{3}{4}$? How much greater?

   (b) Which distance is greater: $\frac{7}{8}$ or $\frac{3}{4}$? How much greater?

   (c) What must be added to $\frac{1}{4}$ to make $\frac{1}{3}$?

   (d) What must be added to $\frac{1}{3}$ to get $\frac{5}{12}$?

TEACHING NOTE:

A **Fraction Fringe** can be cut with the Ellison Letter Machine™ that represents 1, $\frac{1}{2}$, $\frac{1}{4}$, $\frac{1}{8}$, and $\frac{1}{16}$ [there is also a fringe for 1, $\frac{1}{3}$, $\frac{1}{6}$, $\frac{1}{9}$, and $\frac{1}{12}$]. The one pictured connects nicely to measuring to the nearer half-inch, quarter-inch, eighth-inch, and sixteenth-inch with a ruler. It can be used to demonstrate equivalent fractions as well as addition and subtraction of fractions with denominators of 2, 4, 8, and 16. You can compare fractions with the Fraction Fringe. When cutting the fringe, it is useful to keep the colors for fraction parts consistent with colors for the Circle and Square Fraction Models.

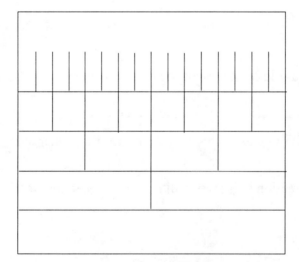

(1 = blue, $\frac{1}{2}$ = green, $\frac{1}{4}$ = orange, $\frac{1}{8}$ = yellow, and $\frac{1}{16}$ = magenta)

Activity 6-11: Mix It Up!

Overview: Students need experience with fractions which are not proper: $\frac{3}{3}, \frac{12}{4}, \frac{5}{2}, \frac{11}{6}$.

Remember that a proper fraction is one in which the numerator is less than the denominator. If a fraction does not fit the definition of proper fraction, we say it is an improper fraction. [Note: improper does not mean, here, "not good." It means "does not fit the definition of proper fraction."] It is important to model, draw pictures, and write symbolically what was done physically *before* developing a rule.

Work with a partner.

Materials: Circle or Square Fraction Models (Appendix Pages 8 & 9).

A. One person takes twelve third pieces and can only use those pieces. The other takes whole units as well as third pieces and makes fair trades when possible. Each person starts with one-third in his work space and will orally name the amount as a one-third piece is added repeatedly.

	one-third	two-thirds	three-thirds	four-thirds	five-thirds
Partner A:					
Partner B:					
	one-third	two-thirds	one	one and one-third	one and two-thirds

Symbolically write each common fraction greater than $\frac{2}{3}$ as an equivalent whole number or mixed number.

B. Using the circle or square fraction model, put three halves in your work space. Make a fair trade of 2 halves for 1 whole.

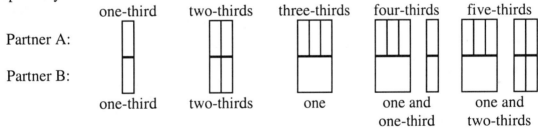

Model each of the following common fractions; make fair trades of fraction pieces for units. Draw a picture to show the process and write each fraction as an equivalent whole or mixed number.

(a) $\frac{7}{5} =$

(b) $\frac{8}{4} =$

(c) $\frac{11}{3} =$

(d) $\frac{13}{6} =$

C. Model each mixed number. Trade each unit for the fraction pieces indicated. Write each mixed number as a common fraction. An example is shown below.

$$2\tfrac{1}{3} = 2 + \tfrac{1}{3} \qquad\qquad \tfrac{6}{3} + \tfrac{1}{3} = \tfrac{7}{3}$$

(a) $4\tfrac{1}{2} = \dfrac{}{2}$

(b) $1\tfrac{1}{6} = \dfrac{}{6}$

(c) $2\tfrac{3}{4} = \dfrac{}{4}$

D. Do you remember what we learned to "recite" if we wanted to change $3\tfrac{1}{5}$ to fifths? "Five times three is 15; plus one is 16 and put the 16 over 5 to get $\tfrac{16}{5}$." WHY do we *say* this? Consider the model:

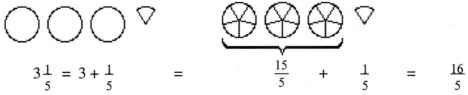

$$3\tfrac{1}{5} = 3 + \tfrac{1}{5} \qquad = \qquad \tfrac{15}{5} + \tfrac{1}{5} = \tfrac{16}{5}$$

In our "rule," when we multiplied 5×3, we were really finding the number of fifths in 3 units since each unit is 5 fifths. Then we added the one (fifth) to the 15 (fifths) to get 16 (fifths). The rule, just recited to students, won't make sense unless the modeling and connecting is done with fraction pieces and symbols. So, $6\tfrac{2}{3} = 6 + \tfrac{2}{3} = \tfrac{18}{3} + \tfrac{2}{3} = \tfrac{20}{3}$ makes sense.

Explain to a partner how to write $5\tfrac{3}{4}$ as a common fraction as if you were using the fraction pieces. Write the number sentence and draw pictures of the fraction pieces to accompany your explanation.

Activity 6-12: Readiness for Multiplying with Fractions

Overview: Readiness activities for multiplication with fractions can be begun using a discrete set model for fractions as soon as students know that taking $\frac{1}{2}$ of a set means dividing that set into two equal-sized parts. To shade $\frac{3}{4}$ of a set of twelve objects, follow these steps: (1) divide the set into 4 equal groups and then (2) shade 3 of the 4 equal groups. So, $\frac{3}{4}$ of 12 = 9.

Materials: Paper; colored pencils; colored chips, squares, or counters

A. Color each set below according to the fraction.

(a)

Color $\frac{3}{4}$ red.

$\frac{3}{4}$ of 8 = _____

(b)

Color $\frac{2}{3}$ orange.

$\frac{2}{3}$ of 6 = _____

(c)

Color $\frac{2}{5}$ blue.

$\frac{2}{5}$ of 10 = _____

(d)

Color $\frac{2}{3}$ yellow.

$\frac{2}{3}$ of 12 = _____

B. Using colored counters, model each of the following.

(a) With 10 counters, make $\frac{2}{5}$ red and $\frac{3}{5}$ yellow.

(b) With 16 counters, make $\frac{3}{8}$ red and $\frac{5}{8}$ yellow.

(c) With 12 counters, make $\frac{3}{6}$ red, $\frac{2}{6}$ blue, and $\frac{1}{6}$ yellow.

Activity 6-13: Multiplying with Fractions

Overview: The multiplication of a whole number and a fraction should be connected to students' previous knowledge of multiplication of whole numbers as repeated addition. The process should include modeling with manipulatives, the use of the semi-concrete number line, and symbolically writing the problem as repeated addition.

Materials: Fraction pieces from the Circle Fraction Model (Appendix Page 8) or other fraction sets

A. $3 \times \frac{1}{5}$ means 3 groups of $\frac{1}{5}$ $\implies$ 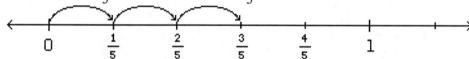 $= \frac{3}{5}$

On a number line, $3 \times \frac{1}{5}$ means 3 "jumps" of $\frac{1}{5}$ and is pictured as

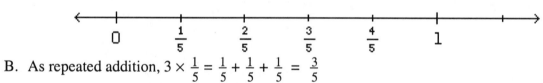

(a) Model $2 \times \frac{2}{5}$ using fraction pieces to determine the product.

(b) Model $2 \times \frac{2}{5}$ on a number line to determine the product.

![number line from 0 to 1 with marks at 1/5, 2/5, 3/5, 4/5]

B. As repeated addition, $3 \times \frac{1}{5} = \frac{1}{5} + \frac{1}{5} + \frac{1}{5} = \frac{3}{5}$

(a) Write $2 \times \frac{2}{5}$ as a repeated addition problem to determine the product.

(b) Complete the chart by writing each multiplication as repeated addition and computing.

Problem	Repeated Addition	Answer
$3 \times \frac{1}{5}$	$\frac{1}{5} + \frac{1}{5} + \frac{1}{5}$	$\frac{3}{5}$
$2 \times \frac{2}{5}$		$\frac{4}{5}$
$5 \times \frac{3}{4}$		
$4 \times \frac{2}{3}$		
$3 \times \frac{2}{11}$		

C. Look for a pattern and generalize a rule for multiplying a whole number and a common fraction.

Activity 6-14: Dividing with Fractions (I)

Overview: This activity uses the partition division model of dividing a fraction into some number of equal-sized parts. The divisor must be a counting number. $\frac{4}{5} \div 2$ means, "Separate $\frac{4}{5}$ into 2 equal-sized parts." How much is in each part? $\left(\frac{2}{5}\right)$

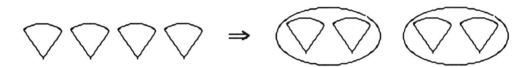

Materials: Fraction pieces from Circle Fraction Model (Appendix Page 8) or other fraction sets

Model each of the division problems with manipulatives, draw pictures to indicate your work and record the answer. Fair trading may be necessary.

(a) $\frac{9}{8} \div 3 =$

(b) $\frac{8}{5} \div 4 =$

(c) $\frac{3}{4} \div 2 =$

(d) $\frac{3}{2} \div 6 =$

(e) $\frac{5}{6} \div 2 =$

Activity 6-15: Dividing with Fractions (II)

Overview: This activity uses the measurement division model and leads to the "common denominator" algorithm for the division of fractions. $2 \div \frac{2}{3}$ means "How many groups of $\frac{2}{3}$ are in 2?" and $\frac{2}{3} \div \frac{1}{6}$ means "How many groups of $\frac{1}{6}$ are in $\frac{2}{3}$?" or just "How many $\frac{1}{6}$'s are in $\frac{2}{3}$?"

Materials: Fraction pieces from Circle Fraction Model (Appendix Page 8) or other fraction set

A. Complete each sentence; use a fraction model to solve the problem.

(a) $\frac{1}{4} \div \frac{1}{8}$ means "How many $\frac{1}{8}$'s are in $\frac{1}{4}$?"

$\frac{1}{4} \div \frac{1}{8} =$ _____

(Note that you "covered" the $\frac{1}{4}$ piece with eighths so that your problem became, $\frac{2}{8} \div \frac{1}{8}$.)

(b) $\frac{3}{4} \div \frac{2}{8}$ means _____

$\frac{3}{4} \div \frac{2}{8} =$ _____

(c) $\frac{3}{2} \div \frac{3}{4}$ means _____

$\frac{3}{2} \div \frac{3}{4} =$ _____

(d) $\frac{9}{2} \div \frac{3}{2}$ means _____

$\frac{9}{2} \div \frac{3}{2} =$ _____

(e) $\frac{12}{5} \div \frac{3}{5}$ means _____

$\frac{12}{5} \div \frac{3}{5} =$ _____

(f) $\frac{15}{6} \div \frac{5}{6}$ means _____

$\frac{15}{6} \div \frac{5}{6} =$ _____

B. Using your work from Part A, complete (a) through (f) below:

(a) $\frac{2}{8} \div \frac{1}{8} =$ _____

(b) $\frac{6}{8} \div \frac{2}{8} =$ _____

(c) $\frac{6}{4} \div \frac{3}{4} =$ _____

(d) $\frac{9}{2} \div \frac{3}{2} =$ _____

(e) $\frac{12}{5} \div \frac{3}{5} =$ _____

(f) $\frac{15}{6} \div \frac{5}{6} =$ _____

C. Generalize a rule for dividing fractions with like denominators.   This rule is called the "common denominator algorithm" for dividing fractions.

TEACHING NOTE:

It is vital for students to understand why the "invert-and-multiply" method works. For example,

$$\frac{2}{3} \div \frac{5}{7} = \frac{\frac{2}{3}}{\frac{5}{7}} = \frac{\frac{2}{3}}{\frac{5}{7}} \times \frac{\frac{7}{5}}{\frac{7}{5}} = \frac{\frac{2}{3} \times \frac{7}{5}}{1} = \frac{2}{3} \times \frac{7}{5}$$

so we can say: "To divide by a fraction, we can multiply by the reciprocal of the divisor." Thus, $\frac{2}{3} \div \frac{5}{7} = \frac{2}{3} \times \frac{7}{5} = \frac{14}{15}$. If we use the common denominator algorithm, we would write:

$$\frac{2}{3} \div \frac{5}{7} = \frac{14}{21} \div \frac{15}{21} = 14 \div 15 = \frac{14}{15} .$$

Activity 6-16: Fractions with Cuisenaire® Rods

Overview: Fraction concepts using a length model can be developed with Cuisenaire® Rods (or with centimeter grid strips).

Materials: Cuisenaire® Rods (can be cut with the Ellison Letter Machine™ dies or can use centimeter grid strips cut from centimeter grid paper on Appendix Page 26) See Activity 4-13 for description of Cuisenaire® Rods.

A. (a) Find all one-color trains that are the same length as the dark green rod. Record in the space provided.

dark green	
green	green

   (b) If the dark green rod is 1, what is the value of the green rod? _____

   (c) Which rod is $\frac{1}{3}$ of the dark green rod? _____

   (d) What fraction of the dark green rod does the white rod represent? _____

B. (a) Cut a 12 cm strip [1 cm × 12 cm] and make all one-color trains the same length as the 12 cm strip.

   (b) Which rod represents $\frac{1}{2}$ of the strip? _____

   (c) Which rod represents $\frac{1}{3}$ of the strip? _____

   (d) $\frac{1}{4}$ of the strip? _____ $\frac{1}{6}$ of the strip? _____ $\frac{1}{12}$ of the strip? _____

   (e) Using the rods, explain why $\frac{1}{2} = \frac{2}{4} = \frac{3}{6} = \frac{6}{12}$.

C. (a) What fraction of the purple rod is the red rod? _____

   (b) What fraction of the dark green rod is the red rod? _____

   (c) What fraction of the brown rod is the red rod? _____

   (d) What fraction of the orange rod is the red rod? _____

(e) Explain why the red rod represents different fractions in each of the previous instances.

D. If the 12 cm strip is 1, you can find the sum of $\frac{1}{2}$ and $\frac{1}{6}$ using the rods. Make a train using the dark green rod and the red rod. Then make one-color trains of the same length.

dark green			red
red	red	red	red
white white white white	white white	white white	
purple		purple	

$\frac{1}{2} + \frac{1}{6} = \frac{4}{6}$ (using red rods)

$\frac{1}{2} + \frac{1}{6} = \frac{8}{12}$ (using white rods)

$\frac{1}{2} + \frac{1}{6} = \frac{2}{3}$ (using purple rods)

Using the 12 cm strip as 1, find $\frac{1}{3} + \frac{1}{6}$ using the rods.

E. With a partner, determine each of the following using the rods. Take the 12 cm strip as 1. Explain your reasoning. Problem (a) is a partitive division problem (Activity 6-14); (b) and (c) are measurement division problems (Activity 6-15); and (d) is comparison subtraction.

(a) $\frac{5}{6} \div 2$

(b) $\frac{5}{6} \div \frac{1}{12}$

(c) $\frac{2}{3} \div \frac{1}{6}$

(d) $\frac{3}{4} - \frac{1}{3}$

(e) $4 \times \frac{1}{6}$

Activity 6-17: Writing Fractions

Overview: Fraction concepts for the unit region area model are first developed where the unit region is divided into congruent parts. At about the fifth-grade level, fraction parts can also be thought of as parts of equal area and then as the part of the region that is shaded. Remember that the area of a triangle = $\frac{1}{2}$ × base × height.

A. The obtuse scalene triangle to the right has been divided into two parts. Justify that each part represents one-half.

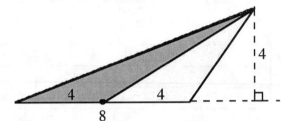

B. In the triangle to the right, show thirds. Justify your answer.

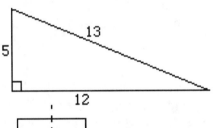

C. Write a fraction for the part of each region that is shaded. The square is the unit region.

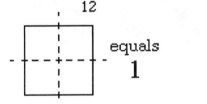

(a)
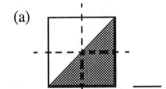

(b)
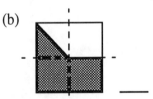

(c)
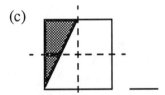

(d)
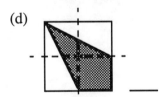

(e)
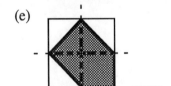

(f)
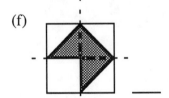

(g)
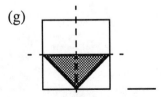

(h)
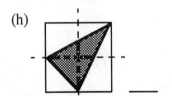

(i)
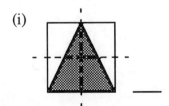

Activity 6-18: Fractions as Ratios

Overview: This activity connects the discrete set model of fractions to ratios as sets of two different colors of counters are compared. If you count out 3 yellow counters and 2 blue counters, there are 5 counters altogether. From knowledge of the discrete set model for fractions, we say $\frac{3}{5}$ of the set is yellow and $\frac{2}{5}$ of the set is blue. However, when we compare the set of 3 yellow counters to the set of 2 blue counters, we say they are in the ratio of 3 to 2. We generally write a "ratio of 3 to 2" as "3 : 2" or as "$\frac{3}{2}$" or sometimes as "3 ÷ 2."

Materials: Various colors of counters.

A. Form a group of 12 counters that has 5 red and 7 yellow.

Write the ratio of red counters to yellow counters in at least three ways: _____, _____, _____.

What fraction of the 12 counters is red?

B. Form a group of counters that has 4 blue and 3 orange counters.

Write the ratio of blue counters to orange counters in at least three ways: _____, _____, _____.

What fraction of the group of counters is blue?

C. The ratio of red to yellow counters is 2 to 3. If there are 5 counters in all, how many are red? _____ yellow? _____ What fraction of the counters is red? _____ If there are 10 counters altogether, how many are red? _____ yellow? _____ What fraction of the counters is red? ___ If there are 15 counters altogether, how many are red? _____ yellow? _____ What fraction of the counters is red?_____

D. In your class of students, what is the ratio of students wearing socks to students not wearing socks? _____ What is the ratio of males to females? _____ What is the ratio of those who wear glasses to those who do not? _____ What is the ratio of students to desks? _____ What is the ratio of students who have never been on a space shuttle mission to those who have been on a space shuttle mission? _____

E. Are there any of the problems in (d) for which you are unable to write a meaningful ratio? Why? What conditions must hold for a ratio comparing the number of elements in two sets to be meaningful?

Activity 6-19: Multiplying Decimals

Overview: This activity connects the concept of area (cover and count) and the rectangular array area model for multiplication (Activity 4-10).

A. This 10×10 grid is a square with an area of 1 square unit, and the length of each edge is 1 unit.

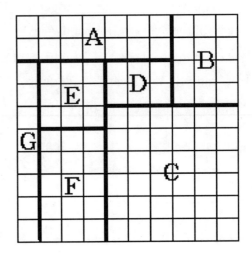

What are the dimensions of each small square? (Write as a decimal.)

As a decimal, write the area of each small square:

B. Recall that, in the rectangular array area model for multiplication, the area of the rectangle is the product of the two dimensions. Write the dimensions and area of each rectangle.

Rectangle	Dimensions	Area
A	$.2 \times .7$	.14
B	___ × ___	
C	___ × ___	
D	___ × ___	
E	___ × ___	
F	___ × ___	
G	___ × ___	

C. Write each decimal numeral in the product as a fraction and multiply. Change each answer to a decimal numeral. Complete the table.

Factors as decimals	Factors as fractions	Product as fraction	Product as decimal
$.4 \times .3$	$\frac{4}{10} \times \frac{3}{10}$	$\frac{12}{100}$	.12
$.2 \times .4$	$\frac{2}{10} \times \frac{4}{10}$	$\frac{8}{100}$	.08
$.14 \times .3$	$\frac{14}{100} \times \frac{3}{10}$	$\frac{42}{1000}$	.042
$.8 \times .9$			
$.26 \times .7$			
$.08 \times .12$			
$.34 \times .07$			
$.002 \times .3$			
$.123 \times .04$			

D. Look at your results from Part C and complete the table.

		# of digits to the right of decimal point		
Factors	Product	first factor	other factor	product
$.4 \times .3$	.12			
$.2 \times .4$	.08			
$.14 \times .3$	.042			
$.8 \times .9$				
$.26 \times .7$				
$.08 \times .12$				
$.34 \times .07$				
$.002 \times .3$				
$.123 \times .04$				

E. Look for a pattern in the table in Part D. Generalize a rule for determining the placement of the decimal point in the product of two decimal numerals.

F. Place the decimal point correctly in each product. (You may have to use 0's as place holders.)

(a) $.87 \times .49 =$ 4 2 6 3

(b) $.073 \times .8 =$ 5 8 4

(c) $.389 \times .65 =$ 2 5 2 8 5

(d) $2.4 \times 5.7 =$ 1 3 6 8

(e) $17.03 \times 31.8 =$ 5 4 1 5 5 4

(f) $306.9 \times 411.06 =$ 1 2 6 1 5 4 3 1 4

G. Base ten blocks can be used to model decimal numerals. The flat represents 1, the rod represents .1, and the unit square represents .01. Thus, the number 2 is modeled with 2 flats and the number 1.3 is modeled with 1 flat and 3 rods. Base ten blocks can also be used to demonstrate the product of decimal numbers as a rectangular array area model where the dimensions of the rectangle are the two factors in the product (see Activity 4-10). Two examples follow.

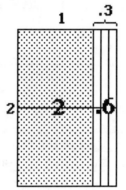

$2 \times 1.3 = 2(1 + .3)$

$= 2 \times 1 + 2 \times .3$

$= 2 + .6$

$= 2.6$

```
ones|tenths
  1 . 3
X     2
   .  6    2 ones × 3 tenths
  2 . 0    2 ones × 1 one
  2 . 6
```

$.4 \times 2.4 = .4(2 + .4)$

$= .4 \times 2 + .4 \times .4$

$= .8 + .16$

$= .96$

```
1's |10ᵗʰ|100ᵗʰ
  2 . 4
X   . 4
   . 1  6   4 tenths × 4 tenths
   . 8      4 tenths × 2 ones
   . 9  6
```

Following this process, use base ten blocks to find the following products.

(a) 4×2.1

(b) $.6 \times 1.7$

(c) 3.1×2.5

(d) $.7 \times 3.1$

TEACHING NOTES: A SUMMARY

Students who have developed an understanding of whole number operations can extend such knowledge to other systems. For example, students need to relate the problem "2.3 + 4.8" to the problem "23 + 48." They should also see how dividing ⁻24 by ⁻3 is like dividing 24 by 3. The concept of multiplication as repeated addition is needed to develop a "rule" for multiplying a whole number by a fraction, but it cannot be used in multiplying two common fractions. Division can be considered as repeated subtraction (measurement division) and this is used in helping develop the common denominator algorithm for dividing two common fractions (Activity 6-15). However, in dividing $\frac{4}{3}$ by 2, partition division must be used. The development of operations on fractions and decimals relies heavily on a student's understanding of varied interpretations of whole number operations. Concrete materials/models help a student lay a strong foundation in developing his/her own understandings. For instance, representational models such as the rectangular array area model can provide a basis for a student's understanding how multiplying 1.2×1.7 is like multiplying 12×17.

The following comments from *Principles and Standards for School Mathematics* give us guidance as we help students from different age groups develop conceptual and computational knowledge of rational numbers.

> In grades 3–5, students' study and use of numbers should be extended to include larger numbers, fractions, and decimals. They need to develop strategies for judging the relative sizes of numbers.

> During grades 3–5, students should build their understanding of fractions as parts of a whole and as division. They will need to see and explore a variety of models of fractions, focusing primarily on familiar fractions such as halves, thirds, fourths, fifths, sixths, eighths, and tenths. By using an area model in which part of a region is shaded, students can see how fractions are related to a unit whole, compare fractional parts of a whole, and find equivalent fractions. They should develop strategies for ordering and comparing fractions, often using benchmarks such as $\frac{1}{2}$ and 1. For example, fifth graders can compare fractions such $\frac{2}{5}$ and $\frac{5}{8}$ by comparing each with $\frac{1}{2}$ -- one is a little less than $\frac{1}{2}$, and the other is a little more.

> In grades 6–8, students should deepen their understanding of fractions, decimals, percents, and integers, and they should become proficient in using them to solve problems. By solving problems that require multiplicative comparisons (e.g., "How many times as many?" or "How many per?"), students will gain extensive experience with ratios, rates, and percents, which helps form a solid foundation

for their understanding of, and facility with, proportionality. The study of rational numbers in the middle grades should build on students' prior knowledge of whole-number concepts and skills and their encounters with fractions, decimals, and percents in lower grades and in everyday life. Students' facility with rational numbers and proportionality can be developed in concert with their study of many topics in the middle-grades curriculum. For example, students can use fractions and decimals to report measurements, to compare survey responses from samples of unequal size, to express probabilities, to indicate scale factors for similarity, and to represent constant rate of change in a problem or slope in a graph of a linear function.

In the lower grades, students should have had experience in comparing fractions between 0 and 1 in relation to such benchmarks as 0, $\frac{1}{4}$, $\frac{1}{2}$, $\frac{3}{4}$, and 1. In the middle grades, students should extend this experience to tasks in which they order or compare fractions, which many students find difficult. For example, fewer than one-third of the thirteen-year-old U.S. students tested in the National Assessment of Educational Progress (NAEP) in 1988 correctly chose the largest number from $\frac{3}{4}$, $\frac{9}{16}$, $\frac{5}{8}$, and $\frac{2}{3}$ (Kouba, Carpenter, and Swafford 1989). Students' difficulties with comparison of fractions have also been documented in more recent NAEP administrations (Kouba, Zawojewski, and Strutchens 1997). Visual images of fractions as fraction strips should help many students think flexibly in comparing fractions. As shown in figure 6.2, a student might conclude that $\frac{7}{8}$ is greater than 2/3 because each fraction is exactly "one piece" smaller than 1 and the missing $\frac{1}{8}$ piece is smaller than the missing $\frac{1}{3}$ piece. Students may also be helped by thinking about the relative locations of fractions and decimals on a number line.

RESOURCES

Periodicals:

"A Game Involving Fraction Squares." *Teaching Children Mathematics* 7 (December 2000): 218-222.

Battista, Michael T., ed., Douglas H. Clements, ed., Leslie P. Steffe, prep., and John Olive, prep. "The Problem of Fractions in the Elementary School." *Arithmetic Teacher* 38 (May 1991): 22-24.

Brinker, Laura. "Using Recipes and Ratio Tables to Build on Students' Understanding of Fractions." *Teaching Children Mathematics* 5 (December 1998): 218-224.

Britton, Barbara J., and Sheryl L. Stump. "Unexpected Riches from a Geoboard Quadrilateral Activity." *Mathematics Teaching in the Middle School* 6 (April 2001): 490-493.

Caldwell, Janet. "Communicating about Fractions with Pattern Blocks." *Teaching Children Mathematics* 2 (November 1995): 156-61.

Cramer, Kathleen, and Nadine Bezuk. "Multiplication of Fractions: Teaching for Understanding." *Arithmetic Teacher* 39 (November 1991): 34-37

Curcio, Frances R., and Nadine S. Bezuk, with others. *Understanding Rational Numbers and Proportions. Curriculum and Evaluation Standards for School Mathematics* Addenda Series, Grades 5–8. Reston, Va.: National Council of Teachers of Mathematics, 1994.

"Dominoes as Fractions: Misconceptions and Understandings." *Mathematics Teaching in the Middle School* 5 (November 1999): 162-165.

Edelman, Leslie. "The Fractions of a Day." *Mathematics Teaching in the Middle School* 3 (March 1997): 192-95.

Empson, Susan B. "Equal Sharing and the Roots of Fraction Equivalence." *Teaching Children Mathematics* 7 (March 2001): 421-425.

Esty, Warren W., "The Least Common Denominator." *Arithmetic Teacher* 39 (December 1991): 6-7.

Fey, James T., and Jane Lincoln Miller. "Proportional Reasoning." *Mathematics Teaching in the Middle School* 5 (January 2000): 310-313.

Fields, Wanda M., Robert Glasgow, Gay Ragan, Robert Reys and Deanna Wasman. "The Decimal Dilemma." *Teaching Children Mathematics* 7 (October 2000): 89-93.

"Fractions: What Happens between Kindergarten and the Army?" *Teaching Children Mathematics* 7 (December 2000): 202-206.

Kamii, Constance, and Mary Ann Warrington. "Multiplication with Fractions: A Piagetian, Constructivist Approach." *Mathematics Teaching in the Middle School* 3 (February 1997): 339-43.

Klein, Paul A. "Remembering How to Read Decimals." *Arithmetic Teacher* 37 (May 1990): 31.

Kouba, Vicky L., Thomas P. Carpenter, and Jane O. Swafford. "Number and Operations." In *Results from the Fourth Mathematics Assessment of the National Assessment of Educational Progress,* edited by Mary Montgomery Lindquist, pp. 64–93. Reston, Va.: National Council of Teachers of Mathematics, 1989.

Kouba, Vicki L., Judith S. Zawojewski, and Marilyn E. Strutchens. "What Do Students Know about Numbers and Operations?" In *Results from the Sixth Mathematics Assessment of the National Assessment of Educational Progress,* edited by Patricia Ann Kenney and Edward A. Silver, pp. 87–140. Reston, Va.: National Council of Teachers of Mathematics, 1997.

Mack, Nancy K. "Building a Foundation for Understanding the Multiplication of Fractions." *Teaching Children Mathematics* 5 (September 1998): 34-38.

Olson, Melfried. "When Will We Reach One-Half?" *Teaching Children Mathematics* 4 (December 1997): 210-14.

Ott, Jack M. "A Unified Approach to Multiplying Fractions." *Arithmetic Teacher* 37 (March 1990): 47-49.

Ott, Jack M., Daniel L. Snook, and Diana L. Gibson. "Understanding Partitive Division of Fractions." *Arithmetic Teacher* 39 (October 1991): 7-11.

Pothier, Yvonne, and Daiyo Sawada. "Partitioning: An Approach to Fractions." *Arithmetic Teacher* 38 (December 1990): 12-16.

Rocke, Judy. "A Common-Cents Approach to Fractions." *Teaching Children Mathematics* 2 (December 1995): 234-36.

Schultz, James E. "Area Models--Spanning the Mathematics of Grades 3-9." *Arithmetic Teacher* 39 (October 1991): 42-46.

Sinicrope, Rose, and Harold W. Mick. "Multiplication of Fractions Through Paper Folding." *Arithmetic Teacher* 40 (October 1992): 116-121.

Slovin, Hannah. "Moving to Proportional Reasoning." *Mathematics Teaching in the Middle School* 6 (September 2000): 58-60.

Sweeney, Elizabeth S., and Robert J. Quinn. " Concentration: Connecting Fractions, Decimals, and Percents." *Mathematics Teaching in the Middle School* 5 (January 2000): 324-328.

Wenzel, Edward J., Cindy L. Anderson, and Kevin M Anderson. "Oil and Water Don't Mix, but They Do Teach Fractions." *Teaching Children Mathematics* 7 (November 2000): 174-178.

Books and Other Literature:

Adler, David A. *Fraction Fun*. New York: Holiday House, 1996.

Creative Publications Staff. *Hands On Pattern Block* (1986). Creative Publications.

Hope, J., B. Reys, and R. Reys. *Mental Math in Junior High* (1988). Dale Seymour Publications.

Hutchins, Pat. *The Doorbell Rang*. New York: Greenwillow, 1986.

Leedy, Loreen. *Fraction Action*. New York: Holiday House, 1994.

Mathews, Louise. *Gator Pie*. Denver, CO: Sundance, 1995.

McMillan, Bruce. *Eating Fractions*. New York: Scholastic, 1991

Murphy, Stuart J. *Give Me Half!* New York: HarperCollins Children's Books, 1996.

Tierney, Cornelia C., and Mary Berle-Carman. *Fractions: Fair Shares. Investigations in Number, Data, and Space*. Palo Alto, Calif.: Dale Seymour Publications, 1995.

Other Resources:

Burns, Marilyn. *Mathematics: With Manipulatives* (1988). Cuisenaire Company of America, Inc. *Pattern Blocks Video*.

7 FROM RATIONAL NUMBERS TO REAL NUMBERS

I hear, and I forget.
I see, and I remember.
I do, and I understand.

Students' first concepts of number and number operations involve the set of whole numbers. As they move through the elementary and into the middle grades, students extend their number sense to include concepts and operations of fractions, decimals, and integers. The use of percents is a natural extension of previous work with fractions and decimals. All these experiences build a knowledge base of rational numbers.

During the middle grades, situations arise for which the set of rational numbers is insufficient. Finding distances using the Pythagorean Theorem, determining the area or perimeter of a circular region, finding the perimeter of a square whose area is 10 square units, or solving a quadratic equation such as $x^2 = 15$ necessitate the use of numbers that are not rational: the irrational numbers. The concept of square root should be developed by using manipulatives to build squares, and students should have opportunities to estimate values of square roots with those manipulatives before using a calculator with the square root key. Students should be allowed to "discover" π (pi) so that they may experience first-hand the joy of such a discovery.

TEACHING NOTE:

Why don't you find a computational answer to the problem **25 ÷ 7** and write it down? Use a calculator, use paper and pencil, or do it in your head. Compare your computational result to the answers you get for the following situations.

(1) I have $25. Each plant in a hanging basket costs $7. How many plants in hanging baskets can I buy for $25?

(2) On a field trip, we have to take 1 adult for every 7 students. If 25 students go on the trip, how many adults have to go?

(3) A trip from Washington to Georgia took 25 days. How many weeks did it take?

(4) A student earned $25 for working 7 hours. How much did the student earn per hour?

Generally, the answer of 3 that you got for (1) and the answer of 4 that you got for (2) are not given as computational answers for **25 ÷ 7**. The mixed number form is required in (3); the decimal approximation is called for in (4).

We must consider problems in CONTEXT. - Peter Hilton

Activity 7-1: Be a Square

Overview: In previous work in Activity 4-10 with the rectangular array area model for multiplication, you saw that when a rectangle is formed by placing squares edge-to-edge, the area is the product of the two dimensions. This activity focuses on special rectangles: squares.

Materials: Squares (color tiles or 1-inch squares cut from the Ellison Letter Machine™) and base ten pieces.

A. The smallest square which can be modeled is the 1×1 square. What is the area of this square? _____ What is the length of an edge of this square? _____

B. Using four unit squares matched edge-to-edge, form a square. What is its area? _____ What is the length of an edge? _____

C. What is the length of an edge of a square made with 9 unit squares matched edge-to-edge? _____ What is the area of this square? _____

D. Let $1 u^2$ be the area of a unit square. Modeling with unit squares, determine:

the length of an edge of a square whose area is $16 u^2$ _____.

the length of an edge of a square whose area is 25 _____.

E. Consider the square to the right. What is the length of an edge of this square whose area is $64 u^2$ _____?

Another way of asking this question is: "What is the square root of 64?" Symbolically, "the square root of 64" is written $\sqrt{64}$. What does $\sqrt{81}$ mean? _____

F. To model $\sqrt{121}$, it is impractical to place 121 unit squares edge-to-edge. Fortunately, the process is simplified by using base ten pieces, as illustrated to the right. To determine $\sqrt{121}$, you would just find the length of an edge of the square whose area is $121 u^2$. $\sqrt{121} =$ _____.

G. Using base ten pieces, form a square whose area is $144 u^2$. $\sqrt{144} =$ _____.

H. Using base ten pieces, form a square whose area is $196 u^2$. (Remember, with base ten pieces, fair trading is allowed.) $\sqrt{196} =$ _____. Repeat to determine $\sqrt{225}$. _____

Activity 7-2: Estimating Square Roots (I)

Overview: Square roots should be estimated with manipulatives before numerical estimation.

Materials: One-inch paper squares of assorted colors, scissors, and base ten pieces.

A. Lay out 5 unit squares. You know that four of them can be arranged into a larger square, with one left over. Therefore, how does $\sqrt{5}$ compare with 2?

The estimate can be made better by cutting the extra square into fourths as shown.

Those four parts can be annexed to the 2×2 square to almost form another square. The edge of this "almost" square measures 2.25 units; there is a hole in the lower, right corner. Because of the missing corner, imagine trimming the same amount off each of the four quarters and combining the trimmings to make the final square. (Note: you will not literally be able to do this!)

Thus, $\sqrt{5}$ is a little less than 2.25.

B. Try a similar process to get an estimate for $\sqrt{6}$. Use paper squares, and then use drawings to explain your work. Thus, an approximate value for $\sqrt{6}$ is _____.

C. $\sqrt{150}$ can be estimated in a similar way. Form 150 using base ten pieces. After fair trading, make a square with an area of 144. Six unit squares are left over. What is the length of each edge of the larger square? _____ (Note that there are 12 units along the bottom and 12 units along each vertical edge.)

Divide the six remaining unit squares into 24 congruent pieces, each of which is one–fourth of a unit. Annex those pieces to the 12×12 square to almost form another square. Since there is an empty spot in the lower right corner, imagine trimming the same amount off each of the 24 quarters and combining to make the final square.

Thus, $\sqrt{150}$ is a little less than _____.

D. Following a similar procedure, estimate $\sqrt{68}$. Show your work below. $\sqrt{68}$ = _____.

E. $\sqrt{150}$ can be approximated a little differently. Form 150 using base ten pieces. Again, after fair trading, make a square with an area of 144. You have 6 unit squares left over.

How many more unit squares must be placed in the shaded region to the right to form the next larger square (13×13)? _____

So, you can see that $\sqrt{150}$ is a little more than 12 but not as much as 13. A better approximation is obtained when an interpretation is given to those 6 squares left over out of the 25 additional squares needed for the next larger square.

How do you write "6 out of 25" as a decimal numeral? _____ We can approximate $\sqrt{150}$ to be about $12 + \frac{6}{25}$. As a decimal numeral, therefore, we can say that $\sqrt{150}$ is approximately _____. (Note: This method is a physical representation of **linear interpolation**.)

F. Approximate $\sqrt{68}$ using the strategy outlined in Part E.

Activity 7-3: Estimating Square Roots (II)

Overview: In previous activities, you estimated $\sqrt{5}$, $\sqrt{6}$, $\sqrt{68}$ and $\sqrt{150}$ using base ten pieces. This activity estimates square roots utilizing an algorithm based on the fact that for a, $b < 1$, $a < b$ if and only if $a^2 < b^2$. Although ultimately students will compute square roots using the square root key on their calculators, this activity helps reinforce the meaning of square root.

Materials: Calculator

A. (a) 5 is between the two squares 4 and 9, *i. e.*, $4 < 5 < 9$. Between which two whole numbers is $\sqrt{5}$? _____ and _____.

   (b) Select a number between these two, such as 2.5 as an estimate of $\sqrt{5}$. By squaring 2.5, you can check your estimate. $(2.5)^2 =$ _____.

   (c) So, 2.5 is too high an estimate. Note that $2.2^2 = 4.84$, so 2.2 is too low an estimate. Also, $2.3^2 = 5.29$, so 2.3 is too high an estimate. This tells you that $\sqrt{5}$ is between 2.2 and 2.3. Furthermore, 2.2 is a closer approximation. Continuing to use this strategy, estimate $\sqrt{5}$ to two decimal places (to the nearer hundredth). $\sqrt{5} \approx$ _____.

   (d) Continuing to use this strategy, estimate $\sqrt{5}$ to three decimal places (to the nearer thousandth). $\sqrt{5} \approx$ _____.

   (e) Check your estimates with the square root key on a calculator.

B. (a) Using the reasoning of Part A, estimate $\sqrt{150}$ to the nearer thousandth. $\sqrt{150} \approx$ _____.

   (b) Compare your estimate to the approximate value you found using the square root key on a calculator.

TEACHING NOTE:

In class one day, a colleague Professor Anne Hudson was asked by a student for a value of $\sqrt{961}$. Another student answered "13." She laughingly said he had reversed the digits as $\sqrt{169} = 13$ but then she recognized that $\sqrt{961} = 31$. As she related this story to me, she also noted that $\sqrt{144} = 12$ and that if the digits in both numbers were reversed, $\sqrt{441} = 21$. We wondered how many more and what kinds of numbers such squares must be. I shared this curiosity with a student teacher who, in turn, shared it with an algebra class. One of her tenth-grade students found over twenty such "reversible squares." Can you find examples? [THERE ARE SO MANY FUN THINGS IN MATHEMATICS THAT CAN COME FROM LISTENING TO STUDENTS!]

There must be an interplay of seriousness and frivolity. The frivolity keeps the reader alert. The seriousness makes the play worthwhile. - Martin Gardner (world's leading recreational mathematician)

Activity 7-4: Where Are Those Square Roots?

Overview: This activity will use the Pythagorean Theorem which is examined in detail in Activities 13-2 and 13-3. You have found that the square roots of most whole numbers are not whole numbers. This activity will give you an idea of where some of these square roots are located on a number line.

Materials: Ruler, 1-inch squares, compass

A. (a) On a piece of paper, draw a line. Label a point near the left end of the line as 0 and give it the name *O*. You will have something like the drawing below.

(b) Place a 1-inch square on the line so that its lower left vertex is at *O* and its lower right vertex is on the line. Mark that point on the line 1 and give it the name *P*. Mark the point at the upper right corner of the square and give it the name *A*. Repeat this process to make tic marks to denote the positions of 2 and 3 on the number line.

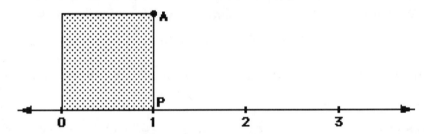

(c) Place the 1-inch square off to the side and connect *O* with *A* and connect *P* with *A*. Your drawing will be like the one below.

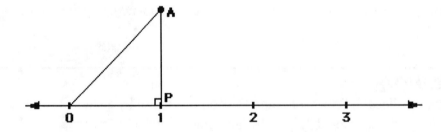

(d) If $\overline{OP}$ is 1 unit long, how long is $\overline{AP}$?

(e) Using the Pythagorean Theorem, find the exact length of $\overline{OA}$.

(f) Using the compass with O as the center, mark an arc from A to the line. Label that point A'. Your drawing will be like the one below.

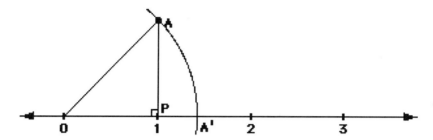

(g) What number on the number line corresponds to the point A'?

B. (a) Place a 1-inch square on $\overline{OA}$ so that its lower right vertex is at A and its lower left vertex is on $\overline{OA}$. Mark the point at the upper right corner of the square and give it the name B.

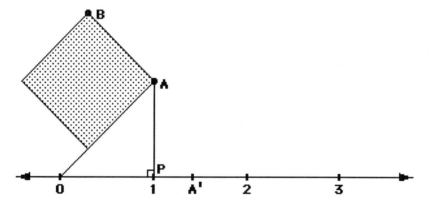

(b) Place the 1-inch square off to the side and connect B with O and connect B with A. Your drawing will be like the one below.

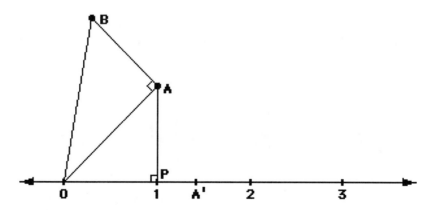

(c) How long is $\overline{OA}$? _____ How long is $\overline{AB}$? _____

(d) Using the Pythagorean Theorem, find the exact length of $\overline{OB}$.

(e) Using the compass with *O* as the center, mark an arc from *B* to the line. Label that point *B′*. Your drawing will be like the one below.

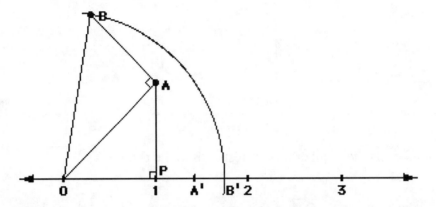

(f) What *number* on the number line corresponds to the point *B′*?

C. (a) Continue as before: place a 1-inch square on $\overline{OB}$ so that its lower right vertex is at *B* and its lower left vertex is on $\overline{OB}$. Mark the point at the upper right corner of the square and give it the name *C*. Place the 1-inch square off to the side and connect *C* with *O* and connect *C* with *B*. Using the Pythagorean Theorem, find the length of $\overline{OC}$.

(b) What is another name for the length of $\overline{OC}$?

(c) Using the compass with *O* as the center, mark an arc from *C* to the line. Label that point *C′*. Your drawing will be like the one below.

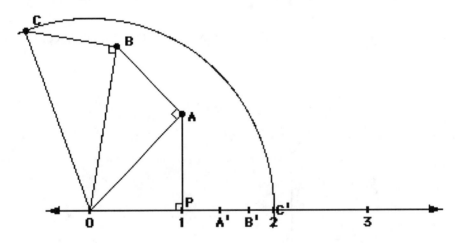

D. Continue in this way to locate $\sqrt{5}$, $\sqrt{6}$, $\sqrt{7}$, and $\sqrt{8}$ on the number line.

Activity 7-5: As American as Apple Pi

Overview: One of the most famous of all irrational numbers is π. The first part of this activity is designed to help you discover the definition of π.
Do this activity in pairs.

Materials: 8 – 15 circular regions of varying sizes (jar lids, pizza boards, cardboard cake rounds from a cake decorating shop, circular fountain on campus, *etc.*), metric tape measures, calculator, and recording sheet

A. Seriate the circular regions by size, label $A - O$. With a partner, measure both the circumference and the diameter of each circular region to the nearer millimeter and record the values in a table like the one below. Both partners should measure and come to consensus as to the closest measure for the circumference and diameter. After measuring and recording, determine the sum, difference, product, and ratio (quotient) of circumference and diameter of each circular region. For convenience, the ratio should be rounded to the nearer thousandth.

Circle	Circumference (C)	Diameter (D)	$C + D$	$C - D$	$C \times D$	$\frac{C}{D}$
A						
B						
C						
•						
•						
•						
N						
O						

B. Discuss with your partner what happens to the quantities in each of the columns "$C + D$," "$C - D$," "$C \times D$," and "$\frac{C}{D}$" as you look at the values which are in those columns from circle A through circle O. Look closely at the values in the "$\frac{C}{D}$" column. Regardless of the size of the circular region, what is the approximate value of $\frac{C}{D}$? _____

This value is close to π: **the ratio of the circumference to the diameter of any circle!**

TEACHING NOTE:

Students often do not realize that π is a constant and will cite its value as exactly 3.14 or $\frac{22}{7}$ even when told these values are only approximations. The "$C + D$," "$C - D$," "$C \times D$" columns were "distractors" in the chart. After completing this activity and discussing the definition of pi (π) as the ratio of the circumference to diameter of any circle, discuss the history of π. In the discussion, include Archimedes' method of using a circle with a diameter of 1 and finding perimeters of inscribed and circumscribed regular polygons ($n = 6, 12, 24, 48, 96$) to approximate π. The video *The Story of Pi* from *Project MATHEMATICS!* by Tom Apostol is a must! You may even want to share a famous cheer. "Three point one four one five nine, block that tackle, hit that line." 3.14159, the first six digits of the number , has long done service as a football cheer at such athletic powerhouses as CalTech. The Greek letter π was not always used to represent the ratio of a circle's circumference and diameter. It was first used to represent this ratio in a book by the English writer William Jones, in 1706; but it was not until 1737 that the current symbol was made popular by the famous Swiss mathematician Leonhard Euler (Beckmann 1971).

Don't forget to celebrate Pi Day on March 14th each year. [At 1:59 p.m. you can eat "pi" without having to worry about the calories!]

Activity 7-6: What Percent Is Shaded?

Overview: Percents are special ratios in which a comparison is made of some number to one hundred. A synonym for *percent* is *hundredth*. The mathematical symbol for percent is % . For example, in the figure to the right, 15 parts of 100 are shaded; this is 15%. This gives a direct connection of percent: first, to a fraction whose denominator is 100 and, secondly, to a decimal numeral. The "cent" in percent has the same meaning as the number of years in a century (100) or the number of cents in one dollar (100). The concept of percent should be developed with models which have already been used with concepts of whole numbers, fractions, and decimals.

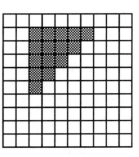

Materials: Transparent 100-grid (Appendix Page 12)

A. Each picture below has 100 equal-sized parts. Write fractions with denominators of 100 and also percents for the shaded part of each picture.

(a) $\frac{7}{100} = 7\%$ (b)   (c)

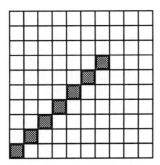 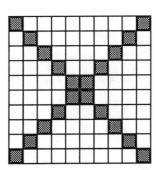 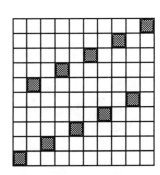

(d)

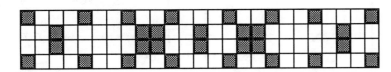

B. Get a partner. Individually, estimate what percent is shaded of each of the squares on the following page. Discuss your estimate with your partner. Then, use a transparent 100-grid to check your estimate. Discuss your strategies for refining your estimates, if needed. Which shaded regions were more difficult to estimate and then refine with the 100-grid? Why?

(a) Estimate ____

Answer ____

(b) Estimate ____

Answer ____

(c) Estimate ____

Answer ____

(d) Estimate ____

Answer ____

(e) Estimate ____

Answer ____

(f) Estimate ____

Answer ____

(g) Estimate ____

Answer ____

(h) Estimate ____

Answer ____

(i) Estimate ____

Answer ____

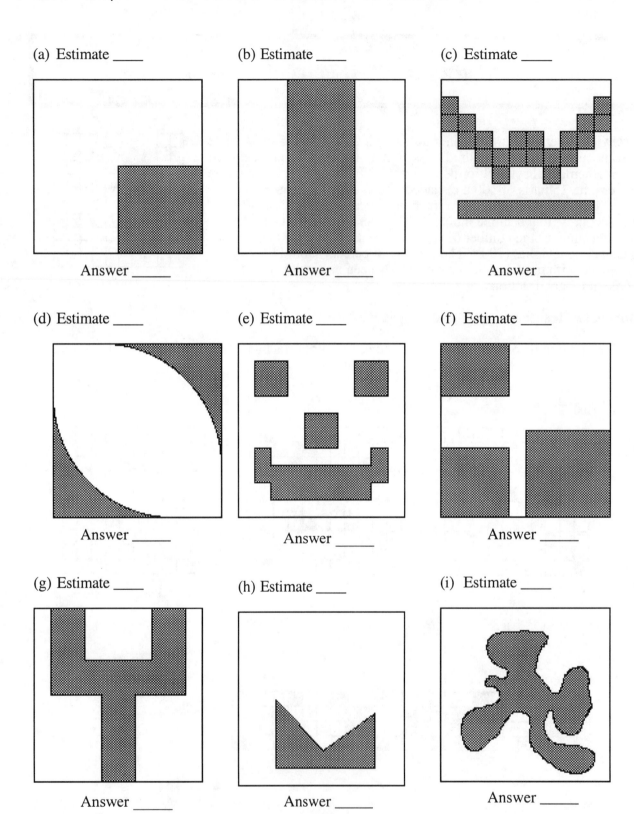

C. Shade to show the correct fraction/decimal/percent. Complete each entry.

Diagram	Fraction	Decimal	Percent
	$\dfrac{17}{100}$		
		.06	
			84%
	$\dfrac{5}{10}$		
			8%

Activity 7-7: Percent: Discrete Model

Overview: Students should work with percents from a discrete set perspective as well as from an area perspective, as many of the real-world percent problems are related to discrete sets.

A. A set of chips is on Timmy's desk, but an index card is covering some of them. 75% of the set is shown. How many chips are in the set?

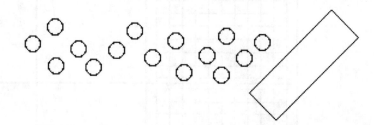

B. This picture represents 125% of a set S. How many triangles are in set S?

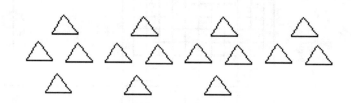

C. There are six chevrons in set C. Below, draw 150% of the set.

D. This picture represents $66\frac{2}{3}\%$ of a set T. How many elements are in set T?

E. Set K is shown to the right.

   (a) How many stars are in set K?

   (b) Below, draw 200% of set K. How many stars are in this set?

   (c) Below, draw 300% of set K. How many stars are in this set?

Activity 7-8: Using a "Percentometer"

Overview: After developing the concept of percent as a part of 100, students can use their knowledge of equivalent fractions to understand how certain percents are convenient fractions. For example, 50% = 50 parts of 100 = 25 parts of 50 = 1 part of 2 = $\frac{1}{2}$. Students should also recognize that $\frac{35}{70}$ is another name for $\frac{1}{2}$ or 50%. Time should be spent allowing students to work with base ten materials so they can understand why $\frac{1}{3} = 33\frac{1}{3}\%$ (and $\frac{2}{3} = 66\frac{2}{3}\%$) or $\frac{1}{8} = 12\frac{1}{2}\%$. Then students are ready to solve problems involving percents using ratio and proportions and a "percentometer."

Materials: Copy of Percentometer (Appendix Page 13)

A. Using your knowledge of equivalent fractions, find the missing value in each percentometer.

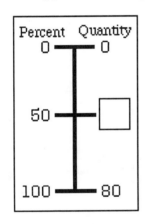 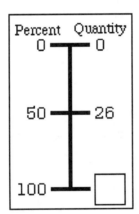 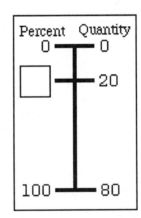 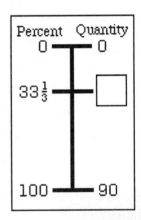

B. To solve percent problems where a basic knowledge of equivalent fractions is insufficient, you can use a proportion equating the ratio on the percent side of the percentometer and the ratio on the quantity side.

For example, let's solve the problem: "A student got 70 of 80 problems correct on a test. What percent of the problems did he get correct?" Since 100% of a quantity means all of it, 80 is placed on the Quantity side of the percentometer to the right of 100%. The number 70 should be placed at its approximate position between 0 and 80 on the Quantity side and an indicator segment drawn across the vertical divider. Your drawing should look like the one to the right.

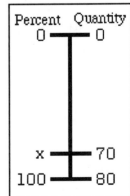

From this drawing, the proportion $\frac{x}{100} = \frac{70}{80}$ follows naturally. When solved, $x = 87.5$. So, 70 out of 80 is 87.5 out of 100 or 87.5%. Use a similar procedure with the percentometer to solve the following problems.

(a) What percent of 60 is 51?

(b) 35% of what number is 80?

(c) 15% of 320 is what number?

(d) 60 is what percent of 72?

(e) 34 is 8% of what number?

C. Just as a thermometer can register a temperature greater than 100°, the percentometer can be easily used in situations where percents exceed 100%. For example, if a student scored 85 points on a test graded out of 80 (he earned bonus points!), he would receive a score of more than 100%.

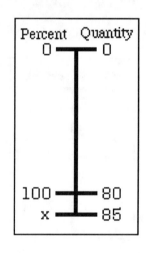

As before, 80 is placed on the Quantity side of the percentometer to the right of 100%. In order to place the number 85 at its approximate position on the Quantity side, however, the percentometer must be extended downward. Once the 85 has been placed, an indicator segment is drawn across the vertical divider. Your drawing should look like the one to the right.

From this drawing, the proportion $\frac{100}{x} = \frac{80}{85}$ follows naturally. Using the notation for percent as parts per 100, you may write $\frac{x}{100} = \frac{85}{80}$. When solved, $x = 106.25$. So, 85 out of 80 is 106.25 out of 100, or 106.25%.

Use a percentometer to solve the following problems.

(a) A normal work week is 40 hours. Last week Justin worked 48 hours. What percent of a full week did he work?

(b) Last year enrollment at Armstrong Atlantic State University was 6,000. This year it is 6,150. This year's enrollment is what percent of last year's enrollment?

(c) Mandy's score on a standardized test is 750. This is 150% of the average score on that test. What is the average score on the test?

> 80% of life is just showing up!
> - Woody Allen

TEACHING NOTES: A SUMMARY

Different representations of numbers are needed or used in different situations. Students should learn to recognize that some representations are much more convenient to use than others. For example, $\frac{1}{3}$ of 30 is somewhat easier to work with than perhaps $33\frac{1}{3}\%$ of 30, and 5.7×10^{-9} is easier to write than the ordinary decimal expansion. The representation of $\frac{25}{99}$ as the repeating decimal .252525. . . is also easier to use in computation. To determine whether students can discriminate between relevant and irrelevant attributes of rational/irrational numbers, they could be asked to determine which of the following represent rational numbers:

$$\frac{5}{7}, \ \sqrt{\frac{3}{5}}, \ 0, \ \sqrt{10}, \ 2.5656, \ ^-4.2, \ 3.020020002\ldots, \ \sqrt{^-25}, \ \frac{14}{16-4^2}, \ \frac{^-20}{^-4}, \ 36\%, \ .\overline{54}$$

From the list above, if a student identifies only $\frac{5}{7}, \ \sqrt{\frac{3}{5}}, \ \frac{14}{16-4^2}, \ \frac{^-20}{^-4}$ as rational numbers (erroneously selecting $\sqrt{\frac{3}{5}}$ as a rational number), he/she may be confusing fraction notation with rational number. If the student identifies 3.020020002 . . . as a rational number, he/she may not be distinguishing repeating decimals (rational) from nonrepeating decimals (irrational) that contain patterns.

It is very important that students develop a solid conceptual and computational understanding of percent. The essential nature of this enterprise is reflected in the following quotes from the NCTM *Principles and Standards for School Mathematics*.

> Students should understand the meaning of a percent as part of a whole and use common percents such as 10 percent, 33 1/3 percent, or 50 percent as benchmarks in interpreting situations they encounter. For example, if a label indicates that 36 percent of a product is water, students can think of this as about a third of the product. By studying fractions, decimals, and percents simultaneously, students can learn to move among equivalent forms, choosing and using an appropriate and convenient form to solve problems and express quantities.

> Percents, which can be thought about in ways that combine aspects of both fractions and decimals, offer students another useful form of rational number. Percents are particularly useful when comparing fractional parts of sets or numbers of unequal size, and they are also frequently encountered in problem-solving situations that arise in everyday life. As with fractions and decimals, conceptual difficulties need to be carefully addressed in instruction. In particular, percents less than 1 percent and greater than 100 percent are often challenging, and most students are likely to benefit from frequent encounters with problems involving percents of these magnitudes in order to develop a solid understanding.

> Students should also develop and adapt procedures for mental calculation and computational estimation with fractions, decimals, and integers. Mental computation and estimation are also useful in many calculations involving percents. Because these methods often require flexibility in moving from one representation

to another, they are useful in deepening students' understanding of rational numbers and helping them think flexibly about these numbers.

RESOURCES

Periodicals:

"Adventures with Sir Cumference: Standard Shapes and Nonstandard Units." *Teaching Children Mathematics* 7 (December 2000): 242-245.

Gerver, Robert. "Discovering Pi - Two Approaches." *Arithmetic Teacher* 37 (April 1990): 18-22.

Haubner, Mary Ann. "Another Look at the Teaching of Percent." *Arithmetic Teacher* 31 (March 1984): 48-49.

Haubner, Mary Ann. "Percents: Developing Meaning Through Models." *Arithmetic Teacher* 40 (December 1992): 232-234.

Lappan, Glenda, James T. Fey, William M. Fitzgerald, Susan N. Friel, and Elizabeth Difanis Phillips. *Comparing and Scaling: Ratio, Proportion, and Percents.* Connected Mathematics series. Palo Alto, Calif.: Dale Seymour Publications, 1998.

Sweeney, Elizabeth S., and Robert J. Quinn. "Concentration: Connecting Fractions, Decimals, and Percents." *Mathematics Teaching in the Middle School* 5 (January 2000): 324-328.

Tent, Margaret W. "Circles and the Number π." *Mathematics Teaching in the Middle School* 6 (April 2001): 452-455, 457.

Thompson, Charles S. and Vicki Walker. "Connecting Decimals and Other Mathematical Content." *Teaching Children Mathematics* 2 (April 1996): 496-502.

Trotter, Terrel. "Let's Take Another Look at Pi Day." *Mathematics Teaching in the Middle School* 7 (March 2002): 374-375.

Books and Other Literature:

Beckmann, Petr. *A History of Pi* (1971). St. Martin's Press.

Neuschwander, Cindy. *Sir Cumference and the Dragon of Pi.* Massachusetts: Charlesbridge Publishing, 1999.

Neuschwander, Cindy. *Sir Cumference and the First Round Table.* Massachusetts: Charlesbridge Publishing, 1997.

Neuschwander, Cindy. *Sir Cumference and the Great Knight of Angleland.* Massachusetts: Charlesbridge Publishing, 1997.

Other Resources:

Apostol, Tom. *The Story of Pi* (1989). *Project MATHEMATICS! Video.*

8 CONSUMER MATHEMATICS

"Well, three or four months run along, and it was well into the winter, now. I had been to school most all the time, and could spell, and read, and write just a little, and could say the multiplication table up to six times seven is thirty-five, and I don't reckon I could ever get any farther than that if I was to live forever. I don't take no stock in mathematics, anyway."

\- Huckleberry Finn

It is somewhat surprising that the mathematics of finance that governs transactions in the marketplace is rooted in concepts as elementary as percent and geometric sequences. In this chapter we focus on activities that make clear this connection. Have your calculator available on each activity and be attentive to the connections between the consumer mathematics and the computational mathematics that you learned earlier.

Activity 8-1: A Matter of Interest

Overview: Simple interest is simply the rent that a borrower pays to use the borrowed money for a period of time. Simple interest is computed as a percent of the amount borrowed. When the borrowed money is kept over several periods, the base on which the percent is computed is kept the same. However, when interest is **compounded** over several periods, the interest is added to the base amount on which the percent is computed at the end of each period. In this activity we will compare and contrast simple and compound interest.

Materials: A calculator

(a) **Simple Interest:** John borrows $3000 from his father to buy a new fishing boat. He agrees to repay the loan at the end of five years and to pay a simple interest rate of 4% a year for the use of the money. Complete the following chart that shows a five-year history of the loan.

Year	Principal	Interest	Amount Owed
1	3000	120	3120
2	3000	120	3240
3	3000		
4	3000		
5	3000		

Notice that the amount owed at the end of five years is:

$$3000 + (3000)(.04)(5) \text{ or}$$
$$3000(1 + (.04)(5)).$$

In general, $A = P(1 + rt)$

where P is the principal or present value, r is the rate paid each year, and t is the time in years.

B. **Compound Interest:** Suppose John agrees to pay his father 4% interest per year compounded annually. In this case, at the end of each year of the loan the interest is added to the principal. Complete the following chart that shows the history of the loan over five years.

Year	Principal	Interest	Amount Owed
1	3000	120.00	3120.00
2	3120	124.80	3244.80
3			
4			
5			

Notice that the amount owed after one year is:
$$3000 + 3000(.04) \text{ or}$$
$$3000(1 + .04) = 3120.$$

Similarly, the amount owed after two years is:
$$3120 + 3120(.04) \text{ or}$$
$$3120(1 + .04) \text{ or}$$
$$3000(1 + .04)(1 + .04) \text{ or}$$
$$3000(1 + .04)^2$$

In general, the compound amount A of a principal (or present value) P at an interest rate of i per period for n periods is given by:
$$A = P(1 + i)^n$$

C. After John repaid the loan, he deposited $800 into savings certificates in preparation for the day when he would need a new motor for his boat. If the savings certificates paid 5% per annum compounded quarterly, then he earned

$$\frac{.05}{4} = \underline{\hspace{2cm}}\%$$

interest for each quarter. Complete the following table that shows the history of this money over the next two years.

Quarter	Principal	Interest	Amount
1	800	10.00	810.00
2	810	10.13	820.13
3			
4			
5			
6			
7			
8			

Notice that the amount in the account after one quarter is:

$$800 + 800(.0125) \text{ or}$$
$$800(1 + .0125).$$

Similarly, the amount in the account after two quarters is $A = 800(1 + .0125)^2$ and the amount after eight quarters is $A = (1 + .0125)^8$.

In general, if an interest rate of r is compounded for k periods each year, the interest rate for each period is $\frac{r}{k}$. Hence, the compound amount of a principal of P compounded k periods a year at an annual rate of r for n years is given by:

$$A = P\left(1 + \frac{r}{k}\right)^{nk}$$

Activity 8-2: From Geometric Sequences to Buying a Boat

Overview: When a sequence of equal payments is made at equal time intervals, the result is called an **annuity**. Often we plan to make purchases at some future time by making regular deposits into an account paying compound interest. In this activity we shall observe that the amount of money in the account at a given time is the sum of a geometric sequence. We shall verify this fact by constructing a period-by-period history of the account.

Materials: A calculator

A. A sequence of the form $a, ar^2, ar^3, ar^4 \ldots$ with $r \neq 0$ is called a **geometric sequence**. In each of the following geometric sequences, identify a, r, and find the next three terms.

(a) 2,   6,   18,   54, _____ , _____ , _____

$a = $ _____ $r = $ _____

(b) 8,   4,   2,   1, _____ , _____ , _____

$a = $ _____ $r = $ _____

(c) 2, $2(1.01)$, $2(1.01)^2$, $2(1.01)^3$, _____ , _____ , _____

$a = $ _____ $r = $ _____

(d) 50, $50(1.0125)$, $50(1.0125)^2$, $50(1.0125)^3$, _____ , _____ , _____

$a = $ _____ $r = $ _____

B. The sum of the first n terms of a geometric sequence: $a + ar + ar^2 + ar^3 + \ldots + ar^{n-1}$
is given by the formula:
$$\frac{a(1-r^n)}{1-r} .$$

Compute the sum of the first 10 terms of each of the geometric sequences in Part A.

C. John decides to begin saving so that he can pay cash on his next purchase of a fishing boat. He decides to deposit $500 at the end of every quarter into an account that pays an annual rate of 4% compounded quarterly. Complete the table below to determine how much he will have at the end of 8 quarters. Notice that the first payment will have been deposited for 7 quarters, so its compound amount is $500(1 + .01)^7$ while his last deposit will have just been made so that its compound amount is $500.

Payment Number	Compound Amount	
1	$500(1 + .01)^7$	536.07
2	$500(1 + .01)^6$	
3		
4		
5		
6		
7		
8	500	

Now add the last column to learn how much John has available to buy his boat after eight quarters of frugality.

D. To compute the amount of money that John had after eight quarters, we need to determine the sum:

$$500 + 500(1.01) + 500(1.01)^2 + 500(1.01)^3 + \ldots + 500(1.01)^7$$

We note that this is the sum of 8 terms of a geometric sequence with

$a = $ _____

$r = $ _____

Compute the sum using the formula for the sum of 8 terms of a geometric sequence and compare the results to your answer from Part C.

(Note: If you rounded your entries in the table in Part C, there will be a two-cent discrepancy between the answer in Part D and the answer in Part C due to rounding error.)

Activity 8-3: From Present Value to Mortgages

Overview: Many of the common financial transactions of the marketplace are based on the arithmetic of geometric sequences. In this activity we will explore the notion of present value and use that idea together with geometric sequences to understand mortgages.

Materials: A calculator

A. Romantic John wishes to buy new fishing gear for his girlfriend on her birthday in six months. How much must he deposit now in an account paying 6% compounded monthly to have $200 available in six months? We know that interest will be compounded at a rate of $\frac{.06}{12}$ = .005 each month and we know that if John deposits P dollars now, it will be worth $P(1.005)^6$ in six months. Hence, we solve the following equation for P :

$$200 = P(1.005)^6$$

$$P = \frac{200}{(1.005)^6}$$

The **present value** of A is the principal P that when compounded for n periods at a rate of i per period will equal the amount A. The present value is computed using the following formula:

$$P = \frac{A}{(1+i)^n}$$

Suppose that money is deposited into an account paying 6% compounded monthly.

(a) What is the present value of $1000 in one year?

(b) What is the present value of $10,000 in five years?

B. When John finally proposed to his girlfriend, he offered to take her fishing in Alaska as a symbol of their engagement, but she was so unreasonable as to insist on an engagement ring. John was short on cash, but arranged to pay the jeweler $200 at the end of each quarter for the next 6 quarters. Money yields a return of 6% compounded quarterly. Complete the table on the following page to compute the cash value of the ring. Observe that the first payment, one quarter from the time of purchase, is presently worth $\frac{200}{1.015}$.

Payment	Computation of Present Value	Present Value
1	$\dfrac{200}{1.015}$	197.04
2		
3		
4		
5		
6		

(a) Add the present values in the last column to determine the cash value of John's ring.

(b) Observe that the numbers that you added in (a) are the terms of a geometric sequence:

$$\frac{200}{1.015} + \frac{200}{(1.015)^2} + \frac{200}{(1.015)^3} + \frac{200}{(1.015)^4} + \frac{200}{(1.015)^5} + \frac{200}{(1.015)^6}$$

Use the formula for the sum of the terms of a geometric sequence to compute this sum.

C. John excitedly called his fiancé to report that he had found the ideal first home. It was a cozy little one-room fishing cabin on the lovely shores of Lake Ouchita. Though a bit rustic, it could be had at the bargain price of $10,000. Though somewhat less enthused, his fiancé inquired about the monthly payment required to obtain this fishy little love nest. John professed ignorance but did report that his Uncle Ned would loan the $10,000 for ten years at 6% interest.

This should be enough information to determine the monthly payment, R. Complete the entries in the following table that should help us understand the present value of this string of 120 payments. Note that the first payment in one month is presently worth $\dfrac{R}{1.005}$ while the last payment is presently worth $\dfrac{R}{(1.005)^{120}}$.

Payment	1	2	3	...	119	120
Present Value	$\dfrac{R}{1.005}$					$\dfrac{R}{(1.005)^{120}}$

Hence, we have the following result:

$$10,000 = R \left(\frac{1}{1.005} + \frac{1}{1.005^2} + \frac{1}{1.005^3} + \ldots + \frac{1}{1.005^{119}} + \frac{1}{1.005^{120}} \right)$$

We can compute R if we can compute this sum

$$\frac{1}{1.005} + \frac{1}{1.005^2} + \frac{1}{1.005^3} + \ldots + \frac{1}{1.005^{119}} + \frac{1}{1.005^{120}}$$

and then divide 10,000 by the sum.

D. The sum $\dfrac{1}{1.005} + \dfrac{1}{1.005^2} + \dfrac{1}{1.005^3} + \ldots + \dfrac{1}{1.005^{119}} + \dfrac{1}{1.005^{120}}$ is the sum of 120 terms of the geometric sequence with

$$a = \frac{1}{1.005} = \underline{\hspace{2cm}}$$

$$r = \frac{1}{1.005} = \underline{\hspace{2cm}}$$

(a) Compute the sum of this geometric sequence. _____

(b) Compute R. _____

(c) What was the monthly payment on John's fishy love nest? _____

TEACHING NOTES: A SUMMARY

Consumer mathematics is introduced to students in the very earliest grades. Early topics focus on how the American monetary system works and the computations needed to make purchases, earn money, and save money (without the complications of interest). As students get older, they are introduced to the ideas of discounts and comparison shopping. By middle school years, they are discussing commissions, sales tax, and simple and compound interest. The topic of annuities discussed in this chapter are not usually found in basic K - 8 curricula, but this topic could certainly be used as an enrichment topic for middle grade students.

RESOURCES

Cohen, Donald. "Can a Purchaser Save Money by Financing?" *Mathematics Teacher* 86 (January 1993): 62-63. *An account of car-salesman trickery.*

Nowlin, Donald. "What Are My Car Payments Going to Be?" *Mathematics Teacher* 86 (April 1993): 299-300. *Computing monthly car payments.*

Thompson, Virginia, and Grace Coates. "Family Math by the Month." *Teaching Children Mathematics* 4 (February 1998): 344-45.

9   INTRODUCTION TO PROBABILITY THEORY

*"In reality no one can teach mathematics. Effective teachers are those who can stimulate students to learn mathematics. Educational research offers compelling evidence that students **learn** mathematics well only when they **construct** their own mathematical understanding. To understand what they learn, they must enact for themselves verbs that permeate the mathematics curriculum: "examine," "represent," "transform," "apply," "prove," "communicate." This happens most readily when students work in groups, engage in discussion, make presentations, and in other ways take charge of their own learning."*

- Everybody Counts

No part of school mathematics is less understood than elementary probability theory. Perhaps this is due to the fact that so often we try to understand probability as an observer rather than as a participant. In the activities of this chapter you will find yourself in the middle of a number of experiments, making observations and trying to describe the likelihood that a specific outcome will occur. This "hands-on" involvement will lead to increased understanding of the basic concepts of probability theory.

Another obstacle to success in understanding probability is the somewhat unusual language that is used. An *experiment* in probability should not evoke the image of a white-coated technician in a laboratory. Rather, an experiment is any process with an uncertain outcome. The set of all possible outcomes of an experiment is called a *sample space*. You should become much more comfortable with these terms as you use them in the context of tossing coins, rolling dice, and spinning spinners.

Despite obstacles, the ability to understand and use elementary probability has grown increasingly important in the last few decades. Whether we wish to model weather or business activity or environmental changes, probability is a primary tool. So, enjoy your work in the activities, get comfortable with the language, and become a more powerful user of probability.

Activity 9-1: Assigning Probabilities

Overview: The probability of an uncertain outcome is a number that measures the likelihood that the outcome will happen the next time the experiment is performed. Probabilities that are assigned on the basis of the properties of the experiment are called **theoretical probabilities**. Probabilities that are assigned after performing the experiment many times are called **empirical probabilities**. In this activity we will compare the two methods of assigning probabilities.

Materials: A paper cup, a thumbtack, a small paper clip, a brown paper bag, five index cards and a copy of the spinner templates on Appendix Page 14. By straightening one end of the paper clip and placing the pencil point through the loop of the clip, one can make a pointer for a spinner as seen below.

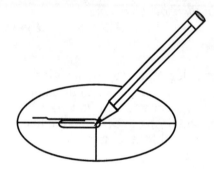

A. Use the paper clip and Circular Region I (Appendix Page 14) to make a spinner.

    (a) One-half of the area of the region is labeled *A*, while *B* and *C* are each used to label one-fourth of the region. From this analysis, assign a theoretical probability to each of the outcomes found in the table below.

A	B	C

    (b) Now, use the "paper clip" spinner and perform this experiment 40 times. Record the number of times that the spinner stops in each region of the circle. Divide each of these three numbers by 40 to assign empirical probabilities.

	A	B	C
Number			
Probability			

    (c) Explain any differences that occurred between the theoretical probabilities and the empirical probabilities.

B. Cut each of the index cards in half to create ten identically sized cards. On three of the cards write the word CAT. On five of the cards write the word DOG. On two of the cards write the word FISH. Place the ten cards in the brown paper bag and shake it thoroughly.

(a) The experiment will consist of drawing a single card from the bag. By thinking about the experiment, assign a theoretical probability to each of the outcomes DOG, CAT, and FISH.

DOG	CAT	FISH

(b) Now repeat the experiment 40 times. Remember after each draw to return the card drawn to the bag and shake the cards thoroughly. Count the number of times that each outcome occurs and then assign empirical probabilities.

	DOG	CAT	FISH
Number			
Probability			

(c) Explain any differences that occurred between the theoretical probabilities and the empirical probabilities.

C. In this activity we will place the thumbtack in the paper cup and shake the thumbtack onto a flat surface. We will record whether the thumbtack lands on its circular base, on its side, or on its point. Since we cannot determine beforehand the theoretical probabilities of this experiment, we will perform the experiment 40 times to determine empirical probabilities.

	⊥	⋏	T
Number			
Probability			

The probability of someone's watching you is directly proportional to the stupidity of your actions. - (unknown)

Activity 9-2: Finding a Uniform Sample Space

Overview: If each outcome in a sample space is equally likely, the sample space is a **uniform sample space**. For a given experiment, there can be many different sample spaces, not all of which are uniform sample spaces. In this activity we will examine two sample spaces for the experiment of flipping a coin twice.
  Work with a partner.

Materials: A pair of nickels and a marker

A. Flip the pair of nickels 50 times. After each flip, determine which of the following three outcomes has occurred: No heads, one head, or two heads. Count the number of times each of the outcomes occurs and record it in the table below.

Outcome	No heads	One head	Two heads
Number			

Does this appear to be a uniform sample space?

B. Use the marker to mark both sides of one of the two nickels with a large *X*. Flip the two nickels 50 times. After each flip determine which of the following outcomes occurred: two heads, *X*-head and tail, head and *X*-tail, two tails. Count the number of times that each of the outcomes occurs and record it in the table below.

Outcome	Two heads	X-head and tail	Head and X-tail	Two tails
Number				

Does this appear to be a uniform sample space?

If not, combine your results with the results of other groups in your class. Now does the sample space appear to be uniform?

(Note: After finishing, clean the *X*'s off the coin because it is illegal to permanently alter United States coins.)

Activity 9-3: Roll a Pair of Dice

Overview: In this activity we will examine two sample spaces for the experiment of rolling a pair of dice.

Materials: Two dice

A. One way to observe the outcomes of this experiment is to observe the sums of numbers showing on the dice. Roll the pair of dice 50 times and count the number of times each outcome occurs. Assign empirical probabilities to each of the outcomes in this sample space.

Sum	2	3	4	5	6	7	8	9	10	11	12
Number											
Probability											

Does it appear that this is a uniform sample space?

B. A second way to observe the outcomes of the experiment of rolling a pair of dice is to observe the number that comes up on each die. For instance, if a 1 shows on the first die and a 2 shows on the second die, the outcome might be recorded as the ordered pair (1, 2). Similarly, the ordered pair (4, 6) would represent a 4 showing on the first die and a 6 showing on the second die. Since each of the possible 36 ordered pairs in this sample space is equally likely to occur, the sample space is uniform and the probability of each outcome is $\frac{1}{36}$. Now let us compare this sample space with the sample space in Part A. For each of the 36 outcomes, record the sum that occurs.

	1	2	3	4	5	6
1						
2						
3						
4						
5						
6						

C. With the help of the table in Part B assign a theoretical probability to each of the outcomes below. For instance, since the ordered pair (1, 1) is the only outcome that yields a sum of 2, the probability assigned is $\frac{1}{36}$. Since both (1, 2) and (2, 1) yield a sum of 3, the probability assigned to 3 is $\frac{2}{36}$.

Sum	2	3	4	5	6	7	8	9	10	11	12
Probability	$\frac{1}{36}$	$\frac{2}{36}$									

How can you account for the differences between these theoretical probabilities and the empirical probabilities in Part A?

TEACHING NOTE:

From the NCTM *Principles and Standards for School Mathematics:*

Ideas about probability at this level [K-2] should be informal and focus on judgments that children make because of their experiences. Activities that underlie experimental probability, such as tossing number cubes or dice, should occur at this level, but the primary purpose for these activities is focused on other strands, such as number.

Students in grades 3–5 should begin to learn about probability as a measurement of the likelihood of events. In previous grades, they will have begun to describe events as certain, likely, or impossible, but now they can begin to learn how to quantify likelihood. For instance, what is the likelihood of seeing a commercial when you turn on the television? To estimate this probability, students could collect data about the number of minutes of commercials in an hour.

Students should also explore probability through experiments that have only a few outcomes, such as using game spinners with certain portions shaded and considering how likely it is that the spinner will land on a particular color. They should come to understand and use 0 to represent the probability of an impossible event and 1 to represent the probability of a certain event, and they should use common fractions to represent the probability of events that are neither certain nor impossible. Through these experiences, students encounter the idea that although they cannot determine an individual outcome, such as which color the spinner will land on next, they can predict the frequency of various outcomes.

Activity 9-4: Is This Game Fair?

Overview: A game of chance is **fair** if each contestant has an equal chance of winning the game. Consider the following two games that involve the use of spinners. After playing them several times, determine if they are fair.
This activity should be completed in groups of two people.

Materials: A small paper clip, a pencil, and the templates for spinners on Appendix Page 14. As in Activity 9-1, we will unfold one end of the paper clip. By placing a pencil point through the loop of the clip onto the center of a spinner template, we will create a spinner.

A. **Game 1**: Player 1 spins Spinner II on Appendix Page 14 and then Player 2 spins the same spinner. If the two players get the same result, then Player 1 wins. If the two players get different results, the Player 2 wins. Choose two contestants, designate one as Player 1 and one as Player 2, and play the game 40 times. Record the results below.

	Player 1	Player 2
Wins		

Does the game appear to be fair? Explain.

B. **Game 2:** In this two-player game, Player 1 spins Spinner II on Appendix Page 14 while Player 2 spins Spinner III on Appendix Page 14. If the sum of the results of the two spins is even, then Player 1 wins. If the sum is odd, then Player 2 wins. Play the game 40 times and record the results.

	Player 1	Player 2
Wins		

Does the game appear to be fair? Explain.

C. (a) Below you will find a tree diagram that describes each of the possible outcomes of Game 1. Some of the branches of the tree have been labeled with their probabilities. Complete the labeling of the tree and then use the probabilities computed from the tree to discuss whether or not Game 1 is fair.

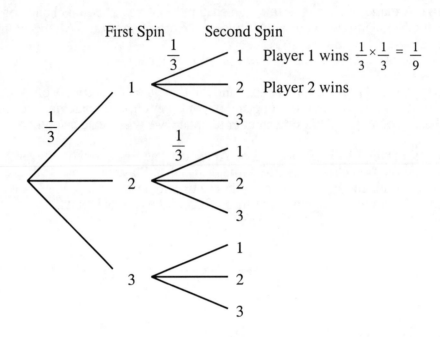

(b) Create a tree diagram showing each of the possible outcomes of Game 2 and use the tree diagram to discuss whether Game 2 is fair.

Activity 9-5: In the Bag!

Overview: A closed bag contains 12 chips—some red, some blue, and some yellow. We are allowed to draw one chip at a time from the bag, but the contents of the bag remain a mystery. If we draw and replace a single chip enough times, we can use our understanding of probability to guess the contents of the bag.
   Work in groups of four on this activity.

Materials: A paper bag and red, blue, and yellow chips. Before starting the activity, one member of the group should secretly place 12 chips in the bag. The bag should either contain

- 6 blue chips, 3 red chips, and 3 yellow chips or
- 2 blue chips, 8 red chips, and 2 yellow chips or
- 1 blue chip, 1 red chip, and 10 yellow chips

   Each of the other three members of the group should perform the following experiment 12 times. Without looking in the bag, reach in and take out a single chip. Record the color, place the chip back in the bag, and shake well to mix the chips. Summarize the results of your work in the chart below.

	Blue	Red	Yellow
Member 1			
Member 2			
Member 3			

What do you believe to be the contents of the bag? Explain!

Activity 9-6: A Simulation: Escape from the Castle of the Iron Knight

Overview: Sir Lancelot escapes from the castle of the Iron Knight and wishes to return to his home in Camelot. The map shows the various routes he can take. However, the Iron Knight occasionally stations soldiers at the bridges along the routes between the castle and Camelot. The probability that a given bridge is open is only $\frac{1}{3}$. We need to determine the probability that Sir Lancelot will be able to find an open route to Camelot.

Materials: A die.

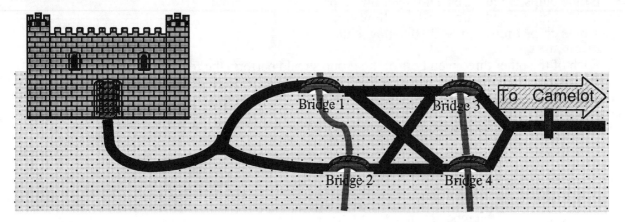

We will simulate the conditions on the day of Lancelot's escape by rolling the die four times. If the first roll shows a 1, 2, 3, or 4, we will conclude that the first bridge is closed; a 5 or 6 will indicate that the first bridge is open. Similarly, the second roll will indicate the conditions at the second bridge and so on. Simulate the experiment 20 times, indicating on each trial whether Lancelot was able to find an open route to Camelot.

Summarize your results in a table like the one found below:

Trial	Bridge 1 open?	Bridge 2 open?	Bridge 3 open?	Bridge 4 open?	Open route?
1	Yes	No	Yes	Yes	Yes
2	No	Yes	No	No	No

Combine the results of your simulation with those of another group in your class. Using these combined results, what is the empirical probability that Sir Lancelot will successfully find a route open to Camelot?

Activity 9-7: Dependent and Independent Events

Overview: Two events in an experiment are said to be **independent** if the occurrence or nonoccurrence of one event has no effect on the probability of the occurrence of the other. If events A and B are independent, then $P(A)$, (the experimental probability of A), in those circumstances in which B has occurred should be about the same as $P(A)$ in all circumstances. Events that are not independent are **dependent**. In this activity we will conduct repeated trials to examine a pair of events that are independent and a pair of events that are dependent.
This activity should be completed in groups of four people.

Materials: For each student in the group: two each of three colors of chips (for example, 2 red (R), 2 yellow (Y), and 2 blue (B)), a bag or box that can be closed, and a colored pencil

A. Each member of the group should perform the following two-step experiment 25 times and record the results:

   (i) Blindly choose a colored chip from the closed box or bag and record its color.
   (ii) Replace the chip, shake the contents and choose a second chip and record its color.

We will be investigating whether the following events are independent:

   • E: The first chip drawn is red.
   • F: The second chip drawn is red.

Trial	1	2	3	4	5	6	7	8	9
Color of Chip 1									
Color of Chip 2									

10	11	12	13	14	15	16	17	18	19	20

21	22	23	24	25

B. (a) Of the 25 times the experiment was conducted, count the number of times that the second chip drawn was red. _____

   (b) Use your colored pencil to circle the columns in which the first chip drawn was red. How many columns did you circle? _____

   (c) Of the columns circled, how many recorded a red chip chosen on the second draw? _____

C. Now combine the information from the work of each of the group members.

   (a) Of the 100 times the experiment was conducted, count the number of times that the second chip drawn was red (the number of times that event F occurred). _____

   $$P(\text{red chip on second draw}) = P(F) = \frac{?}{100} = \underline{\quad}$$

   (b) How many columns were circled by the four group members? _____

   Of the columns circled by the four group members, how many recorded a red chip chosen on the second draw? _____

   $P(\text{red chip on second draw if red chip on the first}) = P(F \text{ occurs if } E \text{ occurs}) =$

   $$\frac{\text{number of times red chip was chosen on both draws}}{\text{number of columns circled}} = \underline{\quad}$$

   (c) Do the results above give evidence that events E and F are independent? Explain.

D. Each member of the group should now perform the following two-step experiment 25 times and record the results:

   (i) Blindly choose a colored chip from the closed box or bag and record its color.
   (ii) *DO NOT REPLACE* the chip, shake the contents, choose a second chip, and record its color.

   We will be investigating whether the following events are independent:

   • E: The first chip drawn is red.
   • F: The second chip drawn is red.

Trial	1	2	3	4	5	6	7	8	9
Color of Chip 1									
Color of Chip 2									

10	11	12	13	14	15	16	17	18	19	20

21	22	23	24	25

E. (a) Of the 25 times the experiment was conducted, count the number of times that the second chip drawn was red. _____

   (b) Use your colored pencil to circle the columns in which the first chip drawn was red:

   How many columns did you circle? _____

   Of the columns circled, how many show a red chip chosen on the second draw? _____

F. Now combine the information from the work of each of the group members.

   (a) Of the 100 times the experiment was conducted, count the number of times that the second chip drawn was red (the number of times that event F occurred). _____

   $$P(\text{red chip on second draw}) = P(F) = \frac{?}{100} = \underline{\quad}$$

   (b) How many columns were circled by the four group members? _____

   Of the columns circled by the four group members, how many show a red chip chosen on the second draw? _____

   $P(\text{red chip on second draw if red chip on the first}) = P(F \text{ occurs if } E \text{ occurs}) =$

   $$\frac{\text{number of times red chip was chosen on both draws}}{\text{number of columns circled}} = \underline{\quad}$$

   (c) Do the results above give evidence that events E and F are dependent? Explain.

TEACHING NOTES: A SUMMARY

Data analysis and probability should be part of childrens' experience at each level of education. Consider the data analysis and probability standard from the NCTM publication *Principles and Standards for School Mathematics* and particularly note the expectations of elementary and middle grades learners related to probability:

DATA ANALYSIS AND PROBABILITY STANDARD

Instructional programs from prekindergarten through grade 12 should enable all students to—

• *Formulate questions that can be addressed with data and collect, organize, and display relevant data to answer them*

• *Select and use appropriate statistical methods to analyze data*

• *Develop and evaluate inferences and predictions that are based on data*

• *Understand and apply basic concepts of probability*

In grades 3–5 all students should–
- describe events as likely or unlikely and discuss the degree of likelihood using such words as *certain, equally likely,* and *impossible;*
- predict the probability of outcomes of simple experiments and test the predictions;
- understand that the measure of the likelihood of an event can be represented by a number from 0 to 1.

In grades 6–8 all students should–
- understand and use appropriate terminology to describe complementary and mutually exclusive events;
- use proportionality and a basic understanding of probability to make and test conjectures about the results of experiments and simulations;
- compute probabilities for simple compound events, using such methods as organized lists, tree diagrams, and area models.

It is particularly important with young children to begin to ask the questions that lead to thinking about processes with uncertain outcomes. Questions such as the following can lead to classroom conversation that opens the door to probability:
- Do you think it will rain today? Do you think we will eat lunch today?
- Do you think that you will you learn anything new today?
- On a die, what numbers are possible to roll? What will you roll?

Children in grades 5 - 8 can be led to understand the importance of probability by helping them find the many references to the topic in newspapers and periodicals. They are acutely interested in fairness, and probability gives an excellent tool for exploring questions of fairness. They are ready for a somewhat more formal study of probability, but the emphasis should continue to be on active exploration of experiments and models.

RESOURCES

Periodicals:

Aspinwall, Leslie, and Kenneth L. Shaw. "Enriching Students' Mathematical Intuitions with Probability Games and Tree Diagrams." *Mathematics Teaching in the Middle School* 6 (December 2000): 214-220.

Austin, Richard A., and Denisse R. Thompson. "*Socrates and the Three Little Pigs*: Connecting Patterns, Counting Trees and Probability." *Mathematics Teaching in the Middle School* 5 (November 1999): 156-161.

"Bingo Games: Turning Student Intuitions into Investigations in Probability and Number Sense." *Mathematics Teacher* 93 (March 2000): 200-206 (see also October 2000, 577).

Brahier, Daniel J. "Genetics as a Context for the Study of Probability." *Mathematics Teaching in the Middle School* 5 (December 1999): 214-221.

Brutlag, Dan. "Choice and Chance in Life: The Game of 'Skunk.'" *Mathematics: Teaching in the Middle School* 1 (April 1994): 28-33.

Edwards, Thomas G. and Sarah M. Hensien. "Using Probability Experiments to Foster Discourse." *Teaching Children Mathematics* 6 (April 2000): 524-529.

"Exploring Probability through an Evens-Odds Dice Game." *Mathematics Teaching in the Middle School* 4 (March 1999): 358-362.

"Free Pizza? Slim Chance!" *Mathematics Teaching in the Middle School* 5 (January 2000): 320-321.

Jardine, Dick. "Looking at Probability through a Historical Lens." *Mathematics Teaching in the Middle School* 6 (September 2000): 50-54.

Jones, Graham A., Cynthia W. Langrall, and Carol A. Thornton. "Using Data to Make Decisions about Chance." *Teaching Children Mathematics* 2 (February 1996): 156-61.

Konold, Clifford. "Teaching Probability through Modeling Real Problems." *Mathematics Teacher* 87 (April 1994): 232-235.

Quinn, Robert J. "Using Attribute Blocks to Develop a Conceptual Understanding of Probability." *Mathematics Teaching in the Middle School* 6 (January 2001): 290-294.

Schwartzman, Steven. "An Unexpected Expected Value." *Mathematics Teacher* 86 (February 1993): 118-120.

"The Probability of Winning a Lotto Jackpot Twice." *Mathematics Teacher* 93 (September 2000): 518-520.

Van Zoest, Laura R., and Rebecca K. Walker. "Racing to Understand Probability." *Mathematics Teaching in the Middle Grades* 3 (October 1997): 33-37.

Books and Other Literature:

Cushman, Jean. *Do You Wanna Bet? Your Chance to Find Out about Probability*. New York: Clarion Books, 1991.

Mori, Tsyoshi and Mitsumasa Anno. *Socrates and the Three Little Pigs*. New York: Philomel Books, 1986.

Newman E., T. Obremski, and R. Scheaffer. *Exploring Probability* (1987). Dale Seymour Publications.

Phillips, Elizabeth, G. Lappan, M. J. Winter, and W. Fitzgerald. *Middle Grades Mathematics Project: Probability* (1986). Addison-Wesley Publishing Company.

Shulte, Albert P., and Stuart A. Choate. *What Are My Chances?* Dale Seymour Publications.

10 | THE USES AND MISUSES OF STATISTICS

"After you understand all about the sun and the stars and the rotation of the earth, you may still miss the radiance of the sunset."
- *Alfred North Whitehead*

The complexities of our society demand that we equip every citizen with the ability to collect, organize, describe, display, and interpret data. A society characterized as an Information Society and based on technology requires more statistics than earlier, less complex social orders. As we prepare to teach statistics, we must be certain that our instruction addresses the skills that are needed in this information society.

A spirit of exploration should permeate the process of learning statistics. Traditionally, the study of statistics in school curricula has been limited to reading graphs and using formulas to compute means or medians. A much more dynamic emphasis is needed. Students should participate in formulating the questions to be asked, collecting the data, and summarizing the data in graphical form, in tables or with measures like mean or median. Beyond this, students should be active in analyzing the results, making conjectures, and communicating the results. Often this discussion will lead to additional conjectures and perhaps the collection of additional data.

In order to get students involved in learning statistics, it is useful to examine data with which they are involved. You will find that in many of the activities of this chapter the data are generated by the participants. In one activity we measure the amount of popcorn that can be held in one hand (and then eat the popcorn); in another we take a physical measurement on all students and use it to draw conclusions. Students respond well to information from surveys that they design (and often learn from the process of writing the surveys). Enjoy doing the activities of this chapter and, as you participate, imagine questions your students might investigate in your own classroom.

Activity 10-1: Choose Your Color

Overview: A crucial first step in understanding and organizing data is the step of seeing the connection between the data and graphical representations of the data. In a **real graph** we use the objects being counted or measured by the data to create the graph, thus facilitating this first step. In this activity we will build two **real graphs** using preferences expressed by the class.
The class will do this activity as a large group activity.

Materials: A set of circular colored chips one inch in diameter (as many of each color as there are students in the class), two small pieces of poster board and a compass to use in drawing a circle (Note: With modifications, one could use Unifix® cubes or colored squares of construction paper in place of the chips.)

A. Pour the multicolored chips into a basket and let each of the students select a chip that represents his/her favorite color. Across the bottom of a small piece of poster board label a column for each color in the basket. Have each student tape his or her chip in the appropriately labeled column as in Figure 1.

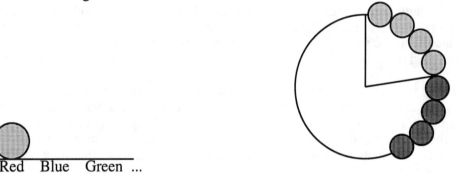

Red   Blue   Green ...

Figure 1   Figure 2

The resulting **real graph** provides a prototype for a bar graph. Use the **real graph** to answer these questions:

(a) Which color was most often chosen? _____

(b) Which color was least often chosen? _____

B. Suppose that there are *n* students in the class. Since each chip is one inch in diameter, the sum of the diameters of the *n* chips is *n* inches. If we wished to draw a circle with a circumference of *n* inches, we would need to remember that circumference is given by $2\pi \times radius$. Thus a circle of circumference *n* inches will have a radius of $\frac{n}{2\pi}$. Replace *n* by the number of students in your class to determine the radius. On a small piece of poster board draw a circle with the appropriate radius. Place the colored chips around the circumference, grouping by color. Now, draw line segments from the center of the circle to the points on the circle where the color of the chips changes. (See Figure 2.)

Activity 10-2: How Sweet It Is: Doing Statistics with M & M´s®

Overview: Statistics is most powerfully used when information from a small subset of a population is used to make predictions about the whole population. Statistics completed with this goal in mind is said to be *inferential statistics.* In this activity we will experience the flavor of *inferential statistics* (in more than one way) as we predict the distribution of colored candy in a bag of M&M's® .

Materials: Snack packs of M&M´s® (better stock up in October!), a two-pound bag of M&M´s® Plain Chocolate Candies, and Appendix Pages 27 and 28. (The two-pound bag of M&M´s® candies should be emptied into a glass jar.).

A. Each bag of M&M´s® Plain Chocolate Candies contains blue, brown, green, orange, red, or yellow candies. Examine the candies in the glass jar and record estimates of the answers to the following questions:

   (a) Which color appears most often in the bag? _____

   (b) What percent of the candies in the bag will be this color? _____

   (c) What color appears least often in the bag? _____

   (d) What percent of the candies in the bag will be this color? _____

B. Estimate the number of M&M´s® in your snack pack before opening. Carefully (that means, without tasting) arrange the candies by color on the diagram in Appendix Page 27. This physical, geometric presentation of the data is called a **real graph**.

C. As you remove the candy from the graph, use the appropriate color to shade the interior of one circle for each M&M®. When completed, you will have a **pictograph** representing these data. In a pictograph, pictures are used to represent the data values being tabulated. In this case each picture of an M&M® represents one piece of candy.

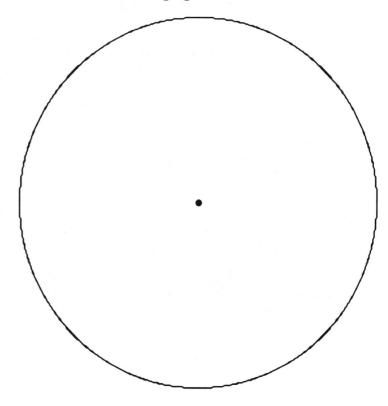

D. Take your M&M´s® and make a circle graph at the right using the method in Activity 10-1. Now, please eat the candy!

Now, use Appendix Page 28 to make a bar graph for your snack pack of M&M's®, coloring in one rectangle for each piece of candy. Compare your circle graph on the previous page to your bar graph. Remember, these two graphs are different ways of displaying the same data.

E. Use the data from your snack pack to complete the following table:

Color	Number	Fractional Part	Decimal Equivalent	Percent
Blue				
Brown				
Green				
Orange				
Red				
Yellow				
Total				

Compare the results from your snack pack with your estimates about the contents of the two-pound bag.

F. Combine the information from your sample with the information from all students in class.

Color	Number	Fractional Part	Decimal Equivalent	Percent
Blue				
Brown				
Green				
Orange				
Red				
Yellow				
Total				

(a) How do the results from the whole class compare with your estimates related to the two-pound bag of candies?

(b) How do the results from the whole class compare with your sample?

(c) How well do you believe the results from the whole class reflect the actual distribution of candies in the two-pound bag?

(d) Make a bar graph depicting the class results on the next page. Then make a circle graph. Remember that a full rotation is 360 degrees and that each color must be represented by the appropriately sized angle.

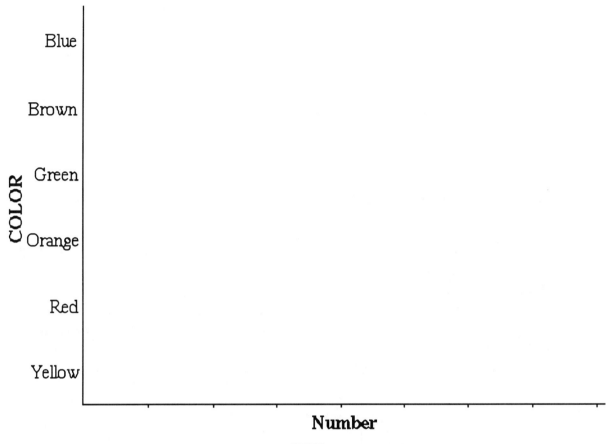

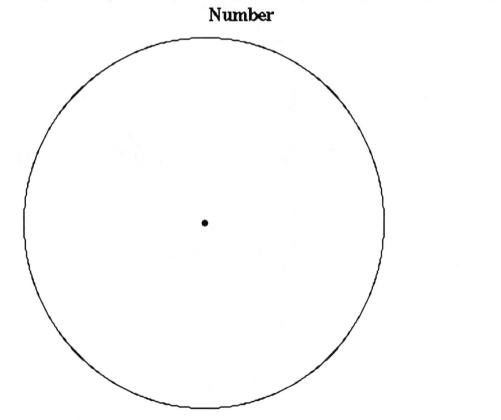

TEACHING NOTE:

When using this activity with children, please note that

- you may have to work with spoonfuls of candy instead of the snack packs of M&M´s®. In this case you will be making inferences about the distribution in a typical spoonful.

- by drawing a circle with an appropriate radius (See Activity 10-1, Part B) you can use the candies in the spoonful to make the circle graph of Part D a real graph.

G. Let *N* be the number of students in your class. Use the information in Part F to create a pictograph for the data from the whole class. (See Part C.) In this case, let each picture of an M&M® represent *N* candies.

Note: Choosing to allow each picture in the pictograph to represent *N* candies, where *N* is the number of students in the class, has two important consequences:

- The pictograph will not be too big.

- The pictograph will represent the distribution of candies in a typical snack pack and will allow us to use the pictograph to draw inferences about the contents of the two-pound bag of candy.

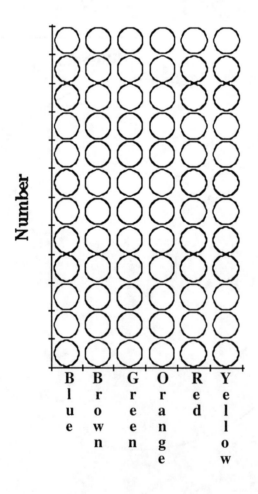

Activity 10-3: Trash Can Olympics

Overview: We form graphs and other visual representations of information in order to be able to see clearly the patterns in the information and to inquire about the source of those patterns. In this activity we perform the same task under several different sets of conditions, create graphical representations of the results of the task, and inquire about possible explanations of what we observe.

Materials: Three trash cans; large, medium, and small; and a ping pong ball for each student

(a) Give each member of the class a ping pong ball. Have each member toss the ball at a large trash can 5 feet from the throwing line. Record the number of balls that stay in the can. Perform this experiment five more times with the indicated modifications.

(b) Form a bar graph with six bars representing the results of the experiment. The length of the bars should represent the number of balls that stayed in the can for each round of the experiment.

Target	Number of Balls
Large can, five feet	
Medium can, five feet	
Small can, five feet	
Large can, ten feet	
Medium can, ten feet	
Small can, ten feet	

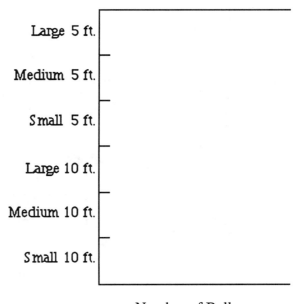

Number of Balls

(c) Explain the results represented by the bar graph.

Activity 10-4: A Handful of Popcorn

Overview: In this activity we will both practice our estimating skills and use statistics to check those skills.

Materials: A large bowl of popped popcorn.

A. Give each member of the class a piece of popcorn and have each one estimate the number of pieces of popcorn that can be scooped and held in one hand. Record these estimates below and then compute the mean of the estimates.

Estimates of Number of Pieces in a Handful of Popcorn

Mean: ____

B. Each member of the class should scoop a handful of popcorn from the bowl. Each student should count the number of pieces of popcorn in his or her hand, record that number, then compute the mean of counts for all the members of the class. Then, eat the popcorn.

Actual Number of Pieces in a Handful of Popcorn

Mean: ____

C. Was the mean of the estimates a good estimate of the mean of actual counts from the class? Why or why not?

Activity 10-5: What's a Cubit?

Overview: A common unit of measure in Biblical times was the cubit, the distance from the end of the elbow to the tips of the fingers. Designate a measuring team for your class and measure the distance to the nearest centimeter from the elbow to the finger tips of each of the students in your class. We will use statistics to analyze what we find.

Materials: Measuring tape, calculator.

A. Record the measurements (elbow to finger tips) for each member of your class. Round to the nearer centimeter.

B. Compute the following statistics for these data:

   (a) Range: _____

   (b) Mode(s): _____

   (c) Median: _____

   (d) Mean: _____

TEACHING NOTE:

Reflect on narrative from the *Principles and Standards for School Mathematics*:

   Much of students' work with data in grades 3–5 should involve comparing related data sets. Noting the similarities and differences between two data sets requires students to become more precise in their descriptions of the data. In this context, students gradually develop the idea of a "typical," or average, value. Building on their informal understanding of "the most" and "the middle," students can learn about three measures of center—mode, median, and, informally, the mean. Students need to learn more than simply how to identify the mode or median in a data set. They need to build an understanding of what, for example, the median tells them about the data, and they need to see this value in the context of other characteristics of the data.

C. Make a grouped frequency distribution for these data, using the following intervals:

Interval	Tally	Frequency
34-36		
37-39		
40-42		
43-45		
46-48		
49-51		
52-54		
55-57		

D. Create a histogram for the grouped frequency distribution in Part C.

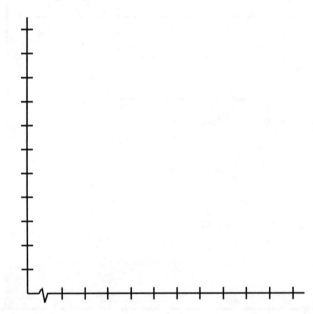

E. What is your evaluation of the cubit as a unit of measurement?

Activity 10-6: Home Run Heroics

Overview: Stem-and-leaf plots and box-and-whisker plots offer two different ways of comparing two sets of data. In this activity we will use both of these devices to compare the home run productions of the two greatest sluggers in baseball history.

A. Below you will find the home run productions for Babe Ruth and Hank Aaron during each of their seasons as major leaguers. Ruth played for 22 years and Aaron played for 23 years.

Season	Ruth	Aaron
1	0	13
2	4	27
3	3	26
4	2	44
5	11	30
6	29	39
7	54	40
8	59	34
9	35	45
10	41	44
11	46	24
12	25	32
13	47	44
14	60	39
15	54	29
16	46	44
17	49	38
18	46	47
19	41	34
20	34	40
21	22	20
22	6	12
23		10
Totals	714	755

Form a stem-and-leaf plot for representing the seasonal home run totals for the two hitters. The stems have been placed in the middle column. Place the leaves for Ruth on the left and the leaves for Aaron on the right.

Ruth   Aaron

| 0 |
| 1 |
| 2 |
| 3 |
| 4 |
| 5 |
| 6 |

B. Rearrange the leaves for each stem in increasing order. Now it should be easy to compute the median number of home runs for both players. Further, if we remember that the first quartile is approximated by the median of the values below the median and the third quartile is approximated by the median of the values above the median, we can compute the first and third quartile for both players.

Ruth ·   Aaron

| 0 |
| 1 |
| 2 |
| 3 |
| 4 |
| 5 |
| 6 |

	Ruth	Aaron
Median	_____	_____
First Quartile	_____	_____
Third Quartile	_____	_____

C. Box-and-whisker plots help us find relationships between two sets of data.

   (a) The fewest number of home runs hit by Ruth was _____

       The greatest number of home runs hit by Ruth was _____

   (b) The fewest number of home runs hit by Aaron was _____

       The greatest number of home runs hit by Aaron was _____

   (c) Locate these points as the ends of the whiskers on the appropriate scale below. Then
       draw a box from the first to third quartiles, draw the whiskers, and indicate the location
       of the median. You have completed the box-and-whisker plots.

Ruth

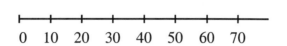

Aaron

D. Use the stem-and-leaf plots and the box-and-whisker plots to make observations about the
home run performances of the two batters.

TEACHING NOTES: A SUMMARY

Topics that involve data collection, data display, and data analysis can be introduced as early as kindergarten. Indeed, Activity 10-1, Choose Your Color, can be used with great success with kindergarten students. When teaching statistics and related topics, focus not only on the creation of the graphs and displays, but also on the interpretation of the data. For instance, in Activity 10-1 you should ask:
- What color is most popular? How do you know this?
- Are you surprised at the results? Explain.
- If you were to ask all the children in the school this same question, would you expect the results to be the same? Explain.

Children of all ages are often quite fascinated with statistics related to measurements taken on members of their class. This interest was utilized in Activity 10-5 as we investigated the length from elbow to finger tip for members of the class. Other interesting questions to examine statistically as a class are: "How long can you hold your breath?" "How far can you throw a softball?" "How tall are you?" "What is the span of your hand from thumb to little finger?" In lower grades it is important to design the experiment so that the child can see the connection between what is being measured and its representation. For instance, one might represent the heights of members in the class by marking the heights of all members of the class on a long strip of adding machine paper. On that strip the mean and median could be found and represented geometrically. Real graphs such as those made in Activities 10-1 and 10-2 are also useful in preserving the connection between data and objects measured.

As children get older, increasing levels of abstraction are possible. Further, their increased knowledge base allows the use of data from a number of different sources. One can illustrate connections between mathematics and science or social studies by using data from these disciplines in descriptive statistics and subsequent analysis and conjecture. Statistics related to sports such as those found in Activity 10-6 will be enthusiastically received by some of your students.

As you study data analysis and statistics with children of different ages, remember the data analysis and probability standard from *Principles and Standards for School Mathematics*:

DATA ANALYSIS AND PROBABILITY STANDARD

Instructional programs from prekindergarten through grade 12 should enable all students to—

- *Formulate questions that can be addressed with data and collect, organize, and display relevant data to answer them*

- *Select and use appropriate statistical methods to analyze data*

- *Develop and evaluate inferences and predictions that are based on data*

- *Understand and apply basic concepts of probability*

An elaboration of each of the standards related to data analysis from the *Principles and Standards for School Mathematics* follows:

- *Formulate questions that can be addressed with data and collect, organize, and display relevant data to answer them*

  In prekindergarten through grade 2 all students should–
  - pose questions and gather data about themselves and their surroundings;
  - sort and classify objects according to their attributes and organize data about the objects;
  - represent data using concrete objects, pictures, and graphs.

  In grades 3–5 all students should–
  - design investigations to address a question and consider how data-collection methods affect the nature of the data set;
  - collect data using observations, surveys, and experiments;
  - represent data using tables and graphs such as line plots, bar graphs, and line graphs;
  - recognize the differences in representing categorical and numerical data.

  In grades 6–8 all students should–
  - formulate questions, design studies, and collect data about a characteristic shared by two populations or different characteristics within one population;
  - select, create, and use appropriate graphical representations of data, including histograms, box plots, and scatterplots.

- *Select and use appropriate statistical methods to analyze data*

  In prekindergarten through grade 2 all students should–
  - describe parts of the data and the set of data as a whole to determine what the data show.

  In grades 3–5 all students should–
  - describe the shape and important features of a set of data and compare related data sets, with an emphasis on how the data are distributed;
  - use measures of center, focusing on the median, and understand what each does and does not indicate about the data set;
  - compare different representations of the same data and evaluate how well each representation shows important aspects of the data.

  In grades 6–8 all students should–
  - find, use, and interpret measures of center and spread, including mean and interquartile range;
  - discuss and understand the correspondence between data sets and their graphical representations, especially histograms, stem-and-leaf plots, box plots, and scatterplots.

- *Develop and evaluate inferences and predictions that are based on data*

  In prekindergarten through grade 2 all students should–
  - discuss events related to students' experiences as likely or unlikely.

  In grades 3–5 all students should–
  - propose and justify conclusions and predictions that are based on data and design studies to further investigate the conclusions or predictions.

  In grades 6–8 all students should–
  - use observations about differences between two or more samples to make conjectures about the populations from which the samples were taken;
  - make conjectures about possible relationships between two characteristics of a sample on the basis of scatterplots of the data and approximate lines of fit;
  - use conjectures to formulate new questions and plan new studies to answer them.

Consider the narrative comments from the *Principles and Standards for School Mathematics*:

> Informal comparing, classifying, and counting activities can provide the mathematical beginnings for developing young learners' understanding of data, analysis of data, and statistics. The types of activities needed and appropriate for kindergartners vary greatly from those for second graders; however, throughout the pre-K–2 years, students should pose questions to investigate, organize the responses, and create representations of their data. Through data investigations, teachers should encourage students to think clearly and to check new ideas against what they already know in order to develop concepts for making informed decisions.
>
> In prekindergarten through grade 2, students will have learned that data can give them information about aspects of their world. They should know how to organize and represent data sets and be able to notice individual aspects of the data—where their own data are on the graph, for instance, or what value occurs most frequently in the data set. In grades 3–5, students should move toward seeing a set of data as a whole, describing its shape, and using statistical characteristics of the data such as range and measures of center to compare data sets. Much of this work emphasizes the comparison of related data sets. As students learn to describe the similarities and differences between data sets, they will have an opportunity to develop clear descriptions of the data and to formulate conclusions and arguments based on the data. They should consider how the data sets they collect are samples from larger populations and should learn how to use language and symbols to describe simple situations involving probability.
>
> Students should become familiar with a variety of representations such as tables, line plots, bar graphs, and line graphs by creating them, watching their teacher create them, and observing those representations found in their environment (e.g., in newspapers, on cereal boxes, etc.). In order to select and interpret appropriate representations, students in grades 3–5 need to understand the nature of different kinds of data: categorical data (data that can be categorized, such as types of lunch foods) and numerical data (data that can be ordered numerically, such as heights of students in a class). Students should examine classifications of categorical data that produce different views. For example, in a study of which cafeteria foods are eaten and which are thrown out, different classifications of the types of foods may highlight different aspects of the data.
>
> Prior to the middle grades, students should have had experiences collecting, organizing, and representing sets of data. They should be facile both with representational tools (such as tables, line plots, bar graphs, and line graphs) and with measures of center and spread (such as median, mode, and range). They should have had experience in using some methods of analyzing information and answering questions, typically about a single population.
>
> In grades 6–8, teachers should build on this base of experience to help students answer more-complex questions, such as those concerning relationships among populations or samples and those about relationships between two variables within one population or sample. Toward this end, new representations should be added to the students' repertoire. Box plots, for example, allow students to compare two or more samples, such as the heights of students in two different classes. Scatterplots allow students to study related pairs of characteristics in one sample, such as height versus arm span among students in one class. In addition, students can use and further develop their emerging understanding of proportionality in various aspects of their study of data and statistics.

RESOURCES

Periodicals:

"Analyzing and Making Sense of Statistics in Newspapers." *The Mathematics Teacher* 92 (April 1999): 318-322.

Bright, George W., Susan N. Friel, and Frances R. Curcio. "Understanding Students' Understanding of Graphs." *Mathematics Teaching in the Middle School* 3 (Nov.- Dec. 1997): 224-27.

"Capture and Recapture Your Students' Interest in Statistics." *Mathematics Teaching in the Middle School* 4 (March 1999): 412-418.

"Collecting Data Outdoors: Making Connections to the Real World." *Teaching Children Mathematics* 6 (September 1999): 8-12.

"Data Analysis and Baseball." *The Mathematics Teacher* 92 (November 1999): 728-732.

Hitch, Chris, and Georganna Armstrong. "Daily Activities for Data Analysis." *Arithmetic Teacher* 41 (January 1994): 242-245.

"How Do Students Think about Statistical Sampling before Instruction?" *Mathematics Teaching in the Middle School* 5 (December 1999): 240-246, 263.

Issacs, Andrew C., and Catherine Randall Kelso. "Pictures, Tables, Graphs, and Questions: Statistical Processes." *Teaching Children Mathematics* 2 (February 1996): 340-45.

Kamii, Constance, Michele Pritchett, and Kristi Nelson. "Fourth Graders Invent Ways of Computing Averages." *Teaching Children Mathematics* 3 (October 1996): 78-82.

Mokros, Jan, and Susan Jo Russell. "What Do Students Understand about Average ." *Teaching Children Mathematics* 2 (February 1996): 360-64.

O'Keefe, James J. "The Human Scatterplot." *Mathematics Teaching in the Middle School* 3 (Nov.-Dec. 1997): 208-209.

Oleson, Vicki L. "*Incredible Comparisons*: Experiences with Data Collection." *Teaching Children Mathematics* 5 (September 1998): 12-16.

Parker, Janet, ed., and Connie C. Widmer, ed. "Statistics and Graphing." *Arithmetic Teacher* 39 (April 1992): 48-52.

Passannante, Marian R. C., and Linda M. Russo. "Statistics Fever." *Mathematics Teaching in the Middle School* 6 (February 2001): 370-376.

Petraroja, Byron, and Thomas R. Scavo. "Adventures in Statistics." *Teaching Children Mathematics* 4 (March 1998): 394-400.

Shaughnessy, J. Michael, and Judith S. Zawojewski. "Mean and Median: Are They Really So Easy?" *Mathematics Teaching in the Middle School* 5 (March 2000): 436-440.

"Statistics in Context." *Mathematics Teacher* 93 (January 2000): 54-58.

"Studying Proportions Using the Capture-Recapture Method." *The Mathematics Teacher* 92 (March 1999): 215-218.

Uccellini, John C. "Teaching the Mean Meaningfully." *Mathematics Teaching in the Middle School* 2 (November–December 1996): 112–15.

Books and Other Literature:

Burrill, Gail F., Jack C. Burrill, Patrick W. Hopfensperger, and James M. Landwehr. *Exploring Regression: Data-Driven Mathematics.* White Plains, N.Y.: Dale Seymour Publications, 1999.

Clement, Jennie, Erin DiPerna, Jean Gavin, Sally Hanner, Eric James, Angie Putz, Mark Rohlfing, and Susan Wainwright. *Children's Work with Data.* Madison, Wis.: Wisconsin Center for Education Research, University of Wisconsin—Madison, 1997.

University of North Carolina Mathematics and Science Education Network. *Teach-Stat Activities: Statistical Investigations for Grades 3 through 6.* Palo Alto, Calif.: Dale Seymour Publications, 1997.

Zawojewski, Judith S. *Dealing with Data and Chance.* Dale Seymour Publications.

11 INFORMAL GEOMETRY

Geometry is grasping space . . . that space in which the child lives, breathes and moves. The space that the child must learn to know, explore, conquer, in order to live, breathe and move better in it.
- Frudenthal

Students are able to represent and make sense of our world through geometry. Students need spatial understanding to interpret, understand, and appreciate our inherently geometric world. Spatial understandings can be developed by providing students opportunities to visualize, draw, measure, build, and explore relationships among geometric entities. Because their spatial abilities often exceed their numerical skills, children's interest in mathematics can be encouraged if those strengths are tapped. Conceptual understandings of number and numerical operations are enhanced with geometric models. For example, you have already used base ten pieces, a proportional manipulative that is a geometric model for whole numbers. Additionally, you have used an area model to interpret algorithms for the multiplication of whole numbers as well as much of the arithmetic of fractions, decimals, and percents. Geometry can be a great motivator in the classroom as children are allowed to investigate, experiment, and explore it with everyday objects and with mathematics manipulatives.

Too often, early geometry experiences have dealt with the memorization of words and ideas that may not make sense to students. Elementary geometry should be informal, allowing time for students to construct their own geometric understandings. The use of physical models will help develop a student's ability to visualize relationships of objects in spatial settings. Spatial thinking is required in tasks such as map reading, understanding diagrams and illustrations, and in putting together things around the house like tables or tents or toys. Basic ideas of symmetry and proportion are found in our homes and in our clothing, in science and in architecture, in art and in seashells.

In addition to its utility in exploring and understanding other areas of mathematics, geometry is closely associated with other subjects, such as art, science, and social studies. For example, students' work on symmetry can enhance their creation and appreciation of art, and their work on coordinate geometry is related to the maps they create or use in their study of the world. The study of geometry promotes a deeper understanding of many aspects of mathematics, improves students' abstract reasoning, and highlights relationships between mathematics and the sciences.
- Principles and Standards of School Mathematics

Activity 11-1: Analyzing Angles

Overview: Students' first exposure to angles should be the *square corner* (or right angle) formed at the corner of a page in a book or the corner of an index card or the corner of the room. The term "right angle" comes from a Greek word meaning "to walk upright." Three angles found in convex polygons should be introduced early on: a square corner or right angle, an acute angle, and an obtuse angle. The meanings of acute and obtuse angles are developed by comparing the angles to a square corner. An acute angle is *sharper* or smaller than a square corner. An obtuse angle is *thicker* or larger than a square corner. To make the comparison, slide the base of an index card along the side of an angle until the corner of the card is on top of the vertex of the angle.

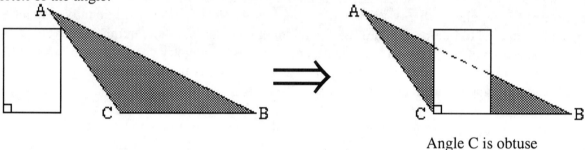

Angle C is obtuse

Materials: Index card

Using the convex polygons below, take the square corner of an index card to use for comparison as indicated above. Identify each angle as right, acute, or obtuse.

Angle	Acute, Obtuse, Right ?
A	
B	
C	
D	
E	
F	
G	
H	
I	
J	
K	
L	
M	
N	
O	

TEACHING NOTE:

Students will eventually learn the formal definition of an angle as *the union of two rays sharing* a *common endpoint*. They will also learn there are five basic angles:

* a zero angle whose measure is 0°,

* an acute angle whose measure is strictly between 0° and 90°,

* a right angle whose measure is 90°,

* an obtuse angle whose measure is strictly between 90° and 180°, and

* a straight angle whose measure is 180°.

However, in the early grades, angles make sense as "corners." Since these corners or angles are parts of convex polygons, they must have measures that are greater than 0° and less than 180°. When younger students think of an acute angle as one that is smaller than a right angle (or obtuse as greater than the right angle), it's O.K. As these students mature and progress into degree measure of angles, their teachers will help them "extend" their conceptual frameworks to include more precise definitions.

Activity 11-2: Bend an Angle

Overview: In almost all cases, it is easiest to involve students with worksheet- and pencil-based activities. Whenever possible, particularly in the study of geometry, we should allow them to experience the concepts involved using some other media. This simple activity with pipe cleaners will give extra "sticking power" to students' understanding of angle classification.

Materials: Pipe cleaners

A. Form a model of each of the following by bending a pipe cleaner.

    (a) Exactly one right angle
    (b) Exactly two right angles
    (c) Exactly three right angles
    (d) Exactly four right angles
    (e) One obtuse and one acute angle
    (f) Two acute and two obtuse angles
    (g) A straight angle.

B. Using a pair of pipe cleaners, form a model of

    (a) A pair of supplementary angles
    (b) A pair of complementary angles
    (c) A pair of vertical angles
    (d) A pair of adjacent angles.

Activity 11-3: Measuring Angles in Wedges and Half-Wedges

Overview: Prior to having degree measure for angles introduced, students will benefit from measuring angles in nonstandard units. In this activity, we will use "wedges" and "half-wedges" to measure angles. Each wedge is $\frac{1}{12}$ of a circular region. To measure $\angle ABC$ in wedges, place the wedges inside the angle so that vertices lie at the vertex of the angle and the wedges are edge-to-edge.

Materials: Circle Fraction Model $\frac{1}{12}$ pieces (cut using Ellison Letter Machine™ or from Appendix Page 8)

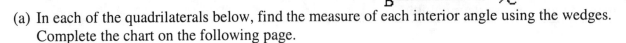

(a) In each of the quadrilaterals below, find the measure of each interior angle using the wedges.
Complete the chart on the following page.

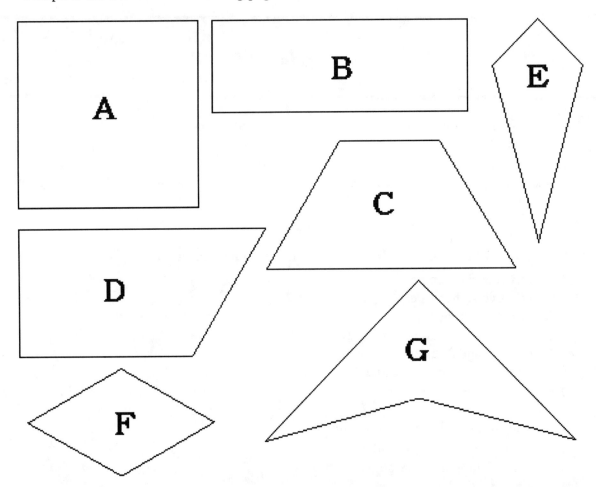

Quadrilateral	A	B	C	D	E	F	G
Sum of measures of angles (in wedges)							

(b) Complete the generalization: If the four angles of a quadrilateral are measured in wedges, then the sum of the measures of the angles is _____.

(c) Cut several of the wedges into halves to make "half-wedges." In each of the triangles below, find the measure of each interior angle using the wedges and half-wedges. Complete the chart.

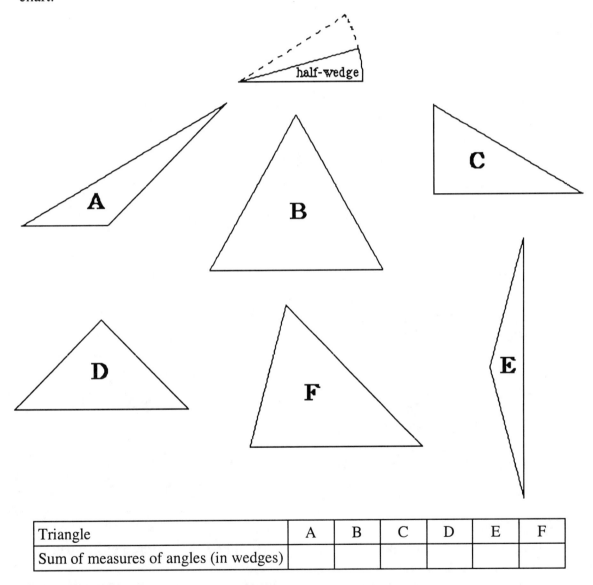

Triangle	A	B	C	D	E	F
Sum of measures of angles (in wedges)						

(d) Complete the generalization: If the three angles of a triangle are measured in wedges, then the sum of the measures of the angles is ____.

Activity 11-4: Making a Wedge/Half-Wedge Protractor

Overview: The Babylonians used a base sixty numeration system. They divided a circle into six sectors and further subdivided each of the six sectors into sixty sectors. The measure of the angle formed by each of the small subdivisions is called 1 degree, or 1°. Thus, one revolution of the circle measures 360° (6 sectors × 60°/sector).

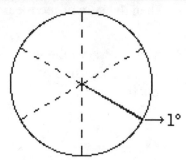

Materials: Wedges and half-wedges cut from $\frac{1}{12}$'s in the Circle Fraction Model (as in Activity 11-3), a $\frac{1}{2}$ circle from black construction paper, glue, scissors, and a protractor

A. Glue six wedges onto paper as shown, forming a semicircular region. Cut out this region. Note that the angle formed by adding the wedges is a straight angle. What is the degree measure of this straight angle?_____ What is the measure, in degrees, of the angle in each of the six wedges? _____

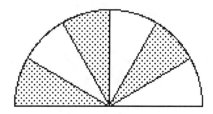

Mark the degree measure on the semicircular region as shown. We will call this device a wedge protractor.

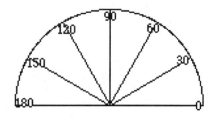

By cutting a wedge in half to form a half-wedge, you have made a sector with a 15° angle. Use a half-wedge to mark off the 15° angles on the wedge protractor as shown.

TEACHING NOTE:

When students have a hand in making a measuring device like a protractor, they can better understand how to use it. Reading a protractor, especially with degree measure scales going in both directions, can be confusing to some students. When measuring angles, only one number on the two scales of a protractor makes sense. Students should be able to determine which of the two measures to use by looking at the size of the angle. It should also be noted that a regular protractor is made by marking each sector with a measure of 1°.

B. Using your wedge protractor, measure each angle below to the nearer 15°.

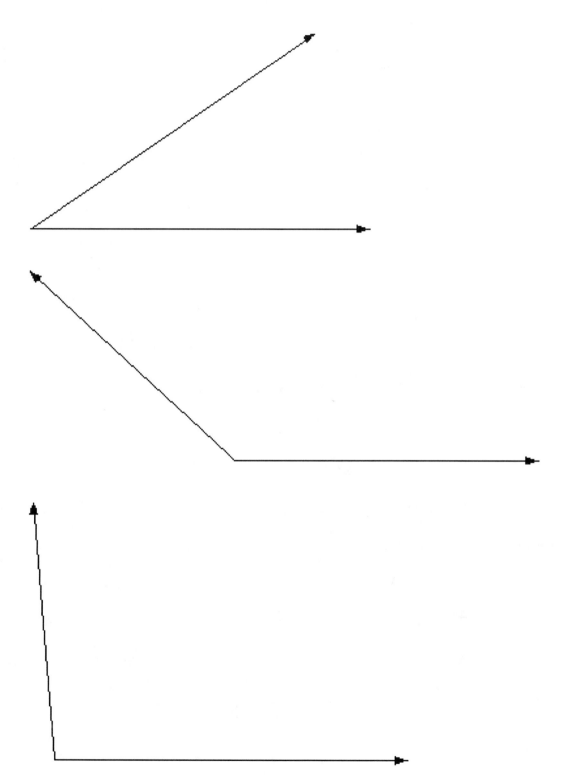

C. Use a commercial protractor to measure each angle to the nearer degree.

Activity 11-5: The Angles of a Triangle

Overview: For those unbelievers from Missouri, this activity offers strong inductive evidence about the sum of the measures of the angles of a triangle.

Materials: Paper, scissors, and protractor

A. Cut several triangles that differ in size and shape. Be sure to include a scalene triangle, a right triangle, an isosceles triangle, and an obtuse triangle in your collection.

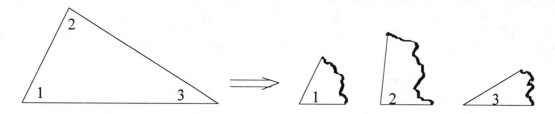

B. Label the angles in each triangle 1, 2, 3. Tear off each angle and add them by placing the vertex of each angle at a point and putting the angles edge-to-edge.

(a) What type of angle is formed in each instance? _____

(b) What is the measure of this angle? _____

(c) What generalization can you make about the sum of the measures of the three angles of a triangle?

C. Another way to illustrate this same generalization is through paper folding. Cut a triangle as indicated below and fold to form an altitude from a vertex. Fold the vertex down to the point of intersection of the altitude with the base of the triangle. Slide the vertex of each base angle along the base until it, too, coincides with the point at which the altitude intersects the base. Note that the three angles of the triangle form a straight angle. Thus, if the sum of the three angles is a straight angle, then the sum of the measures of the three angles of the triangle is 180°.

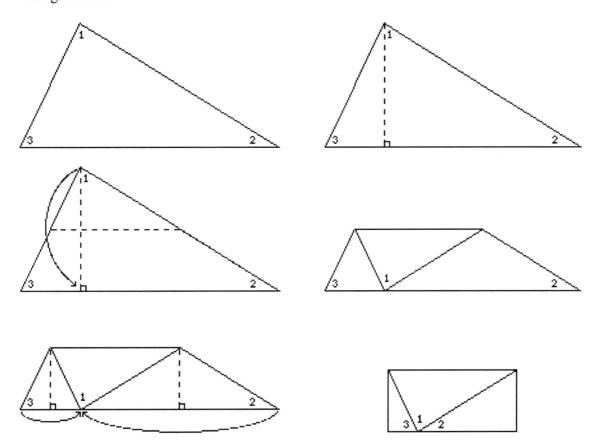

TEACHING NOTE:

Students often think of angles as corners and may have difficulty with "straight" angles. Put two sheets of paper next to each other and talk about the angle made where the two square corners meet. The two right angles, each measuring 90°, form the straight angle which measures 180°. Activities 11-3 and 11-4 measure angles of a triangle in wedges. The angle formed by each wedge ($\frac{1}{12}$ of a circle) is 30° ($\frac{1}{12}$ of 360°). You found that the sum of the measures of the angles in the triangles is six wedges, which measures $6 \times 30°$ or 180°. Likewise, the sum of the measures of quadrilateral angles was twelve wedges, which measures $12 \times 30°$, or 360°.

Activity 11-6: Finding Symmetry Lines

Overview: Symmetry is one of the most fundamental concepts in geometry, introduced to children early in the curriculum. A region is symmetrical if one half of the region can be folded exactly over the other half. In this activity you will sharpen your understanding of symmetry by paper folding. A Mira® can also be used. (See Activities 13-1, 13-12, and 14-3.)

Materials: Tracing paper

Trace each figure below on tracing paper. On each paper shape, estimate and draw in where you think a symmetry fold (a line of symmetry) would be. Check your estimate by folding the shape.

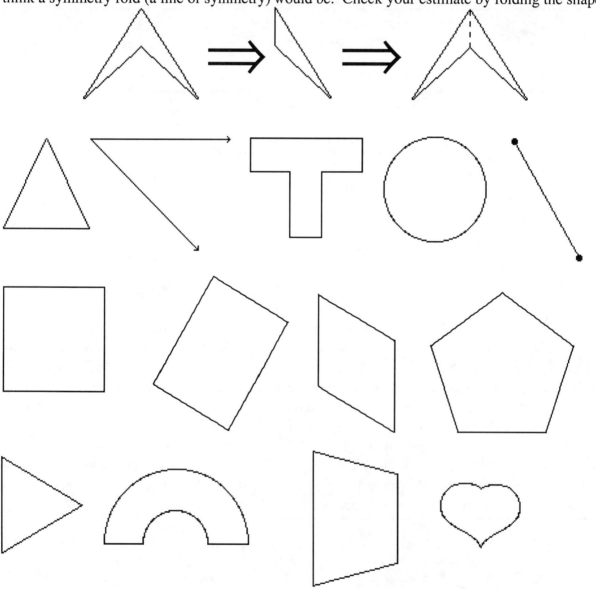

Activity 11-7: Quadrilateral Quandary

Overview: In this activity we will focus on the properties of quadrilaterals that allow us to distinguish one from another.

Materials: A copy of Appendix Page 15

A. On Appendix Page 15 you will find 10 quadrilaterals. Draw the two diagonals in each quadrilateral and, if they intersect, label the points of intersection. Measure each edge, each interior angle, the angles between the diagonals, the diagonals, and the pieces of both diagonals. Label each quadrilateral with your measurements. When the measurements are recorded, complete the following table by checking those properties that hold for each polygon. [Y = yes; N = no.]

	A	B	C	D	E	F	G	H	I	J
(a) All edges congruent	Y	N								
(b) Exactly one pair of parallel edges		Y								
(c) Congruent diagonals										
(d) Two pairs of opposite edges are parallel										
(e) Two pairs of opposite angles are congruent										
(f) Adjacent edges are congruent										
(g) Diagonals are perpendicular										

B. Now complete the following table in which you classify each of the 10 quadrilaterals on Appendix Page 15. (Some of the quadrilaterals may appear on more than one list.)

Classifications	Quadrilaterals
Kite	
Trapezoid	
Parallelogram	A, E
Rhombus	
Rectangle	
Square	

C. Look again at the properties listed in the table from Part A. From your explorations in Part A, complete this table giving those properties that seem to be true of quadrilaterals with each classification.

Classifications	Properties
Kite	
Trapezoid	
Parallelogram	
Rhombus	
Rectangle	
Square	(a)

TEACHING NOTE:
A quadrilateral with congruent edges is often referred to as a diamond, but I often tell children that mature students and mathematicians call it a rhombus. "Oh," but teachers often say, "that word is not in our objectives. It is too hard for younger students." If kindergartners can (and do!) say "Tyrannosaurus Rex" or "brachiosaurus" or "pterodactyl," then it isn't hard for them to say "rhombus." It is important for them to become familiar with mathematical terminology by hearing it as a part of the teacher's everyday language in the classroom.

After studying geometric shapes, one middle schooler said to another: *You're just an equiangular, equilateral quadrilateral – in short, a square.* Not to be outdone, the recipient of this comment retorted: *It's better to be square than to go 'round in the wrong circles!*

Activity 11-8: Triangle Treasure I - Making Shapes from Triangles

Overview: Familiar shapes, such as triangles, can be put together to make new shapes. *Work in pairs.*

Materials: Two congruent triangles of the following types (different colors): scalene right triangles cut from 3 × 5 index cards; isosceles right triangles; equilateral triangles; scalene obtuse triangles

A. Take two scalene right triangles and find all possible planar shapes which can be made by matching the two triangles edge-to-edge. Sketch your solutions. Classify each shape and justify your reasoning. One of the solutions is presented below.

The shape made by the two triangles is an *isosceles obtuse triangle*. Note that in the obtuse triangle the edges opposite congruent acute angles are the hypotenuses of the congruent right triangles and hence those edges are congruent.

B. Find all possible planar shapes which can be made by matching the two congruent isosceles right triangles edge-to-edge. Sketch your solutions. Classify each distinct shape and justify your reasoning.

C. Find all possible planar shapes which can be made by matching the two congruent equilateral triangles edge-to-edge. Sketch your solutions. Classify each distinct shape and justify your reasoning.

D. Repeat the process using two congruent scalene obtuse triangles.

E. The number and types of shapes you made in Parts A, B, C, and D differ. Which type of triangles produced the fewest distinct shapes? _____ Which type produced the greatest number of distinct shapes? _____ Explain why you think they differ.

Activity 11-9: Problem Solving with Pentominoes

Overview: When problem solving is discussed in the elementary curriculum, quite often the attention is given to problems that are verbal in nature. While this attention is quite appropriate and necessary, it misses the fact that there are important spatial problem-solving skills that need to be developed also. This activity with pentominoes and Activities 11-11 and 11-12 with tangrams provide good opportunities for assisting in building those skills.
Work in groups of 2 or 3 persons.

Materials: Tape and one-inch squares cut from multi-colored construction paper (commercially available or easily cut with the Ellison Letter Machine™)

A. **From Dominoes to Pentominoes:** Give each group a large supply of 1-inch squares of various colors.

(a) Make all possible distinct planar shapes that can be made by joining two one-inch squares edge-to-edge. These are called *dominoes*. How many are there?

(b) Make all possible distinct planar shapes that can be made by joining three one-inch squares edge-to-edge. These are called *trominoes*. How many are there? (Be careful, we will regard the two trominoes shown below as the same since they can be made to coincide when placed on top of one another - that is, they are congruent.)

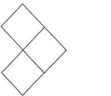

(c) Make all possible distinct planar shapes that can be made by joining four one-inch squares edge-to-edge. These are called *tetrominoes*. How many are there?

(d) Make all possible distinct planar shapes that can be made by joining five one-inch squares edge-to-edge. These are called *pentominoes*. How many are there?

(e) To facilitate problem solving with pentominoes, each of them has a name. Use your imagination (and your fingers to rotate and flip them) and determine which pentomino has each of these twelve one-letter names:

F L I P N T U V W X Y Z

Hint: The **F** and **N** are shown below.

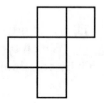
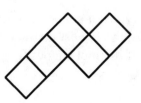

B. **About Pentominoes:** Remember that each pentomino is formed from squares measuring one inch on each edge.

  (a) What is the area of each pentomino? Can you explain why this is the case for all pentominoes?

  (b) Complete the following table giving the perimeter for the pentominoes **F, L, W, X, Y**.

Pentomino	**F**	**L**	**W**	**X**	**Y**
Perimeter	12 in.				

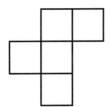

Do you have a conjecture about the perimeter of all pentominoes?

  (c) What is the perimeter of **P**? Does this confirm or disprove the conjecture that you made using inductive reasoning in (b)?

  (d) Each square has a perimeter of 4 inches; hence, the sum of the perimeters of five squares that have not been joined is 20 inches. Each time two squares are joined along an edge in forming a pentomino, two edges are lost from this pool of 20 edges. Use this reasoning to explain why:

  • **F** has a perimeter of 12 inches.

  • **P** has a perimeter of 10 inches.

  • All other pentominoes have perimeters of 12 inches.

(e) Take the **X** pentomino made with your squares and demonstrate that it can be folded and taped to form an open-top cube. Now, without touching the other pentominoes, visually examine them to determine which ones can be folded to form open-top cubes and which cannot. In each case, identify which square of the pentomino is the base. After sorting visually, check your results by folding.

Form open-top cubes	Do not form open-top cubes
X	

C. **Spatial Problem Solving with Pentominoes:** Spatial problem-solving skills can be honed by using a set of pentominoes to form a predetermined figure. (Rich sources of such puzzles can be found in booklets such as *Pentomino Activities* and *Pentomino Lessons* written by Henry Picciotto and published by Creative Publications.)

(a) A rectangle that measures 3 inches by 5 inches has an area of 15 square inches. How many pieces of the pentomino set would it take to fill and completely cover a 3" by 5" rectangle? _____ Find as many different solutions to this problem as you can.

(b) **Challenge Problem:** Draw a 6-inch by 10-inch rectangle. How many pieces of the pentomino set will it take to cover this rectangle? _____ Find a solution (there are over 2,000). One solution can be found on Appendix Page 16.

(c) **Challenge Problem:** Select any pentomino and remove it from the set. Choosing nine of the remaining 11 pentominoes, make a similar shape to the pentomino you have selected and removed. (Note: Each edge of the larger similar shape is three times as long as an edge of the selected pentomino. We say that the ratio of similitude is 3.) What will be the area of the larger similar shape? How does the area of the tripled shape compare with the area of the pentomino you selected?

Activity 11-10: Tiling Patterns

Overview: A rectangle can be used to tile or cover the plane with neither gaps nor overlaps. A regular hexagon like the yellow pattern block piece will also tile with neither gaps nor overlaps as will an equilateral triangle. These shapes are said to **tessellate** because they tile a flat surface in a pattern with neither gaps nor overlaps. Students can explore properties of polygons through tessellations. Students should also use nonconvex polygons like the pentomino shapes (Activity 11-9).

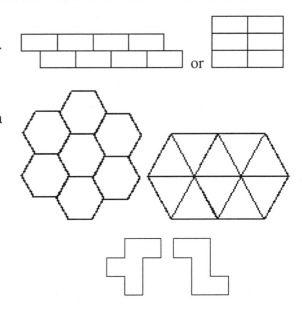

Materials: Red trapezoid and blue rhombus from pattern blocks or pattern block template; 6-8 each of the **F** and **Z** pentomino pieces (Appendix Page 16); centimeter grid paper (Appendix Page 26)

A. Using the red trapezoid, draw two different tessellating patterns.

B. Using the blue rhombus, draw two different tessellating patterns.

C. Make a tessellation using the **F** pentomino and draw your tessellation on centimeter grid paper. Repeat with the **Z** pentomino.

D. Determine which of the other pentomino shapes tessellate: **L I P N T U V W X Y**

TEACHING NOTE:
Tessellations can be found in many familiar places — being aware of them is the key. Look for tessellations on the floors in restaurants, in tiled showers, in stained glass windows in churches, in fabric designs, and in any mall in the tiling designs on floors, walls, or fountains. See also Activity 13-14 and 13-15.

Activity 11-11: Tantalizing Tangrams I - Paper Folding

Overview: Students can develop understandings of geometric manipulatives and their properties when they are allowed to construct them. This activity will allow you, through paper folding, to construct one of the most famous of all geometric puzzles.

Materials: A large paper square with edges of measure at least 8 inches on an edge (nicely cut from $8\frac{1}{2} \times 11$ colored copy paper); scissors

As you fold and cut pieces from your paper square, (*) will denote a finished piece to be put aside.

A. Fold the square region in half along the diagonal and cut along the fold.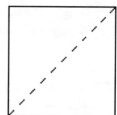

   (a) What is the special name of each of these two shapes?

   (b) What is the degree measure of each of the three angles?

B. Take one of the triangles from (A) and fold in half by matching the two congruent 45° angles. Cut along the fold.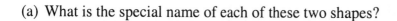

   (a) What is the special name for each of these two shapes?

   (b) Put the two pieces aside.

C. Take the remaining triangle from (A). Fold the hypotenuse in half and crease to locate the midpoint (M) of the hypotenuse. Fold the vertex of the right angle down to that midpoint. Cut along the fold.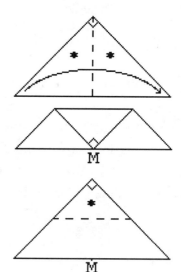

   (a) What are the names of the two shapes?

   (b) Put the triangle aside.

D. Take the remaining trapezoid and fold in half by matching the two acute base angles and the two obtuse base angles. Cut along the fold. This yields two trapezoids resembling baby booties.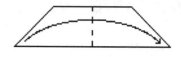

E. Take one of the "booties" from (D) and fold the "heel" of the bootie to the "toe" of the bootie, forming a square and a small isosceles right triangle. Cut along fold and put aside.

F. Take the one remaining bootie from (D) and fold the "heel" across
    to the opposite obtuse vertex. Cut along the fold. What two shapes
    are formed?

The seven pieces that result from this paper folding and cutting are called *tangrams*. You
can clearly see that there are two largest pieces: the large isosceles right triangles. Similarly,
the two smallest pieces are the two small isosceles right triangles.

G. Put the square, the medium triangle, and the parallelogram into your work space. By just
    inspecting these shapes visually, seriate (or order) them by size (that is, determine which one
    looks the smallest, next, and then largest.) After you have made a guess, take the two small
    triangles and cover each shape in turn. Sketch in your solutions below. What does this say
    about the size of these three tangram pieces?

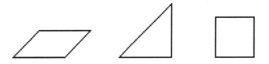

H. Put one of the large triangles into your work space. Cover the large triangle in as many
    different ways as you can with other tangram pieces (but not with itself!). Draw your
    solutions below.

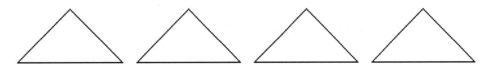

I. You made your seven tangram pieces from a large square. Put the seven tangram pieces
    back together to make that square. Sketch your solution below.

Activity 11-12: Tantalizing Tangrams II - Making Shapes

Overview: Tangram pieces can be used to create many geometric shapes. The activities here are but a few of many challenging tangram puzzles.

Materials: A set of tangrams (cut with Ellison Letter Machine™, cut using Appendix Page 18, or a plastic commercial set)

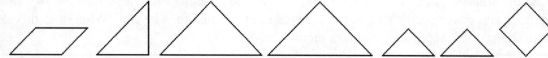

A. You used the seven tangram pieces to form a square in Activity 11-11. Using 2, 3, 4, and then 5 of the tangram pieces, form squares. Draw your solutions below.

B. Using all seven tangram pieces, form a parallelogram which is not a rectangle. Sketch your solution below.

C. Using all seven tangram pieces, form a rectangle which is not a square. Sketch your solution below.

D. **Challenge Problem:** A polygon is **convex** if each segment joining two nonadjacent vertices of the polygon lies inside the polygon. Otherwise the polygon is **nonconvex** (or **concave**). Only 13 convex polygons can be made using all seven tangram pieces. You already know the square. If you completed Parts B and C, you have formed two more. Can you find the rest?

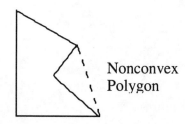

Nonconvex Polygon

E. Create the following shapes using all seven tangram pieces.

TEACHING NOTE:

The tangram puzzle is one of the best loved and most versatile of manipulatives. It originated in China in the early 1800s. Manipulating the shapes will allow your students to discover concepts of size, shape, area, perimeter, similarity, properties of polygons, and even fractions. Note that each of the tangram pieces is a fractional part of the square: the large triangle is $\frac{1}{4}$, the medium triangle is $\frac{1}{8}$, as are both the square and the parallelogram, and each small triangle is $\frac{1}{16}$. Additionally, each piece can be written as a percentage of the original square from which the pieces were cut. Activity 12-13 connects the length of the edge of the square used to fold and cut the tangram pieces to the exact length of the edges of each piece and the sum of their perimeters. Tangrams are also used in developing spatial problem-solving skills and concepts of area.

Activity 11-13: Triangle Treasure II - SuperTangrams™

Overview: If you completed Activity 11-9, you made pentominoes by taping five one-inch squares edge-to-edge. This SuperTangrams™ activity is a similar geometric problem-solving activity and is best done in groups of 3 or 4 to encourage discussion and articulation of reasoning. The shapes you are about to make offer a recreational approach to mathematical concepts.

Materials: 70 - 80 isosceles right triangles cut from multi-colored paper (easily cut with the Ellison Letter Machine™ and Fraction Square $\frac{1}{8}$ B die ⊞); tape

A. Take four isosceles right triangles and make a planar shape by matching the four triangles edge-to-edge (leg-to-leg or hypotenuse-to-hypotenuse). This is one example of such a shape. Working with your group, find all possible distinct planar shapes which can be made by matching four isosceles right triangles together edge-to-edge. Be careful; if two shapes can be made to coincide by placing one on top of the other, they are congruent and are not distinct. Compare the shapes made by your group to those made by other groups.

B. These shapes are called SuperTangrams™ because their underlying geometry is that of the Chinese tangram puzzle. Three of the SuperTangrams™ are identical to three of the tangram pieces. Using your shapes, classify each and sort by the number of edges: triangle, quadrilateral, pentagon, hexagon, etc.

C. Sort the shapes into those which are convex and those which are nonconvex (concave). Recall that a polygon is nonconvex if a segment joining any two vertices lies outside the shape.

   (a) What do you notice about the number of edges of shapes which belong to one set or the other?

   (b) Each isosceles right triangle is a 45°- 45°- 90° triangle. Find the degree measure of the interior angles of each SuperTangrams™ shape. Notice the degree measures of interior angles of convex and then nonconvex SuperTangrams™. In general, how might you define convex and nonconvex polygons in terms of degree measure(s) of interior angles?

D. Since area is a measure of covering, what is an appropriate unit of area for the SuperTangrams™? _____ What is the area of each SuperTangram? _____ Can you explain why this is the case for all of these shapes?

E. Let L represent the length of each leg in one isosceles right triangle, and let H represent the length of its hypotenuse.

(a) Find the perimeter of each SuperTangrams™ shape as an expression in L and/or H.

(b) Sort the SuperTangrams™ by perimeter. Discuss in your group how you could determine the shortest/longest perimeter values. Seriate the perimeters from least to greatest.

(c) If you used the fact that in an isosceles triangle, the hypotenuse length $= \sqrt{2} \times$ leg length, find a way to justify your seriation without using that fact.

F. **Spatial Problem Solving with SuperTangrams™:** Spatial problem-solving skills are sharpened by using SuperTangrams™ to cover/form predetermined shapes. [Rich and challenging puzzles can be found in booklets such as *Supertangram Activities* by Henri Picciotto and published by Creative Publications.] Using two SuperTangrams™, make the shape below in three different ways.

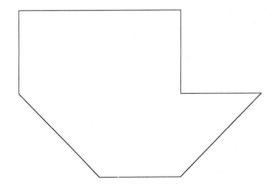

Activity 11-14: Circle Folding

Overview: Just as knowing the vocabulary of a foreign language does not ensure that you can use it in conversation, knowing the vocabulary of geometry does not mean that you can use it. After you complete this activity, find a friend not in the class and lead that friend through the activity by giving verbal directions. This will strengthen your ability to use the vocabulary of geometry.   (This activity was adapted from a session in paper folding led by Evan Maletsky.)

Materials: Two circles with 8" diameters (easily cut with the Ellison Letter Machine™)

A. **Finding the Center:** Place the circular region on the table in front of you and "eyeball" it to determine where the center of the region would be. Mark it with a pencil or pen. Discuss with a partner how you could determine if the point you marked is really the center.

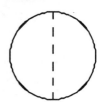

(a) Fold the circular region in half. Is this fold a "line of symmetry"? (Be careful; the fold is actually a segment!)

(b) This symmetry fold is called a diameter. The diameter is also a chord (a segment connecting two distinct points on the circle). Can you make a fold to create a longer chord than the diameter?

(c) Make a new fold to divide the circular region in half. What is this new fold called?

(d) Locate the intersection of the two folds and mark it with your pencil. This point is the center of the circular region. Estimate how many millimeters "off" from the center your estimated center point is (a millimeter is about the width of the wire of a standard paper clip). If you are within 2-3 millimeters, you have good spatial visualization skills!

TEACHING NOTE:

Simple folds of a circular region can be used to illustrate several additional concepts.

- If you folded your circular region in half and then folded the semicircular region in half, your two diameters are perpendicular. What kind of angles are made by these two diameters?

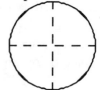

- If you folded your circular region in half, unfolded and then refolded again, your folds may have formed two acute vertical angles and two obtuse vertical angles.

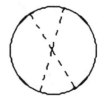

- Fold your circular region in half to "see" the straight angle. Note that the acute angle and the obtuse angle form a linear pair; their sum is a straight angle. What are the angles called when together they form a straight angle?

B. Fold the edge of the circular region down
    so that a point on the circle touches the
    center.

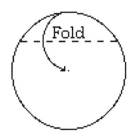 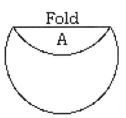

Repeat this, using one endpoint of fold A
as an endpoint of fold B.

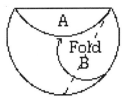 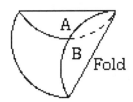

Fold the remaining "flap" down to the center so that the
endpoints of fold A and fold B are the endpoints of fold C.
What kind of triangle is formed by making these folds?

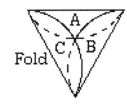

C. Find the midpoint of one edge of the triangle by matching the
    vertices of two of the angles and creasing slightly. Fold the
    opposite vertex to this midpoint. What shape is made by this
    folding?

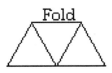

D. Notice that the trapezoid has two congruent acute
    base angles and two congruent obtuse base angles.
    Fold one of the acute angles across to the obtuse
    angle on the opposite edge. What is the special
    name for this new shape?

E. Note that opposite angles in this rhombus are
    congruent. Fold the rhombus in half by matching
    the acute angles.

TEACHING NOTE:

It is beneficial to digress here and examine relationships between the original equilateral
triangle and this small equilateral triangle. Ask the students to consider these questions:
• How does the small equilateral triangle compare to the original equilateral triangle?
• If the edge of the small equilateral triangle is one unit, what is the length of the edge of the
  larger triangle?
• What is the ratio of similitude?
• If the area of the small equilateral triangle is one unit, what is
  the area of the large triangle?
• What is the ratio of the areas of the larger to the smaller
  triangle? (It should be the square of the ratio of similitude.)

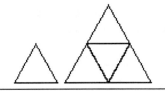

F. Take the small triangle and "open" up the shape, bringing the vertices of the original triangle together to a point to make a three-dimensional surface, a polyhedron.

  (a) What is the name of this polyhedron? _____

  (b) How many faces are there? _____

  (c) What is the shape of each face? _____

  (d) How many vertices are there? _____

  (e) How many edges are there? _____

G. Open the folded shape back up to the large equilateral triangle. Fold each vertex to the center. What shape is formed?

H. Pull the three small triangles which are folded over so that they lie on top of each other and gently push the edges together. You have formed a truncated tetrahedron!

  (a) How many faces are there? _____

  (b) What is the name of this polyhedron? _____

  (c) Describe the two bases.

  (d) Describe the lateral faces.

TEACHING NOTE:

When discussing shapes, students in grades 3–5 should be expanding their mathematical vocabulary by hearing terms used repeatedly in context. As they describe shapes, they should hear, understand, and use mathematical terms such as parallel, perpendicular, face, edge, vertex, angle, trapezoid, prism, and so forth, to communicate geometric ideas with greater precision. For example, as students develop a more sophisticated understanding of how geometric shapes can be the same or different, the everyday meaning of *"same"* is no longer sufficient, and they begin to need words such as congruent and similar to explain their thinking.

  - Principles and Standards of School Mathematics

Activity 11-15: The Shape of Things

Overview: A polyhedron is a simple closed surface and is formed by enclosing a single portion of three-dimensional space with four or more polygonal regions. Each face of a polyhedron is a polygon. An edge is formed when exactly two faces are joined. A vertex is a point where three or more edges meet. If the faces of a polyhedron are congruent regular polygonal regions, and the same number of faces meet at each 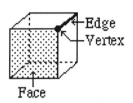 vertex in exactly the same way, then the polyhedron is called a *regular polyhedron*. Since there are an infinite number of regular polygons, you may suppose there are an infinite number of regular polyhedra. However, as Lewis Carroll expressed it, they are "provokingly few in number." There are only five regular convex polyhedra, known as the *Platonic Solids:* (a) tetrahedron, (b) hexahedron (cube), (c) octahedron, (d) dodecahedron, and (e) icosahedron.

Work in pairs.

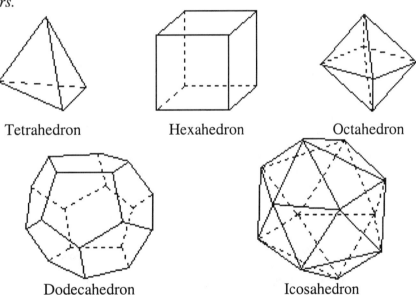

Tetrahedron   Hexahedron   Octahedron

Dodecahedron   Icosahedron

Materials: An assortment of polyhedra, including the Platonic solids; stellated octahedron made from nets in Appendix Pages 19 and 20 [Platonic solids can be made with Polydron™ or from nets, like the ones below, formed by taping together regular polygons (easily cut with the Ellison Letter Machine™)]; *The Platonic Solids* and *Stella Octangula* videos from Key Curriculum Press

(a)   (b)   (c) 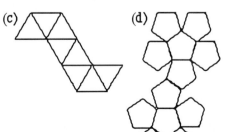 (d) (e)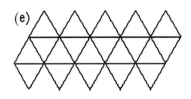

A. Determine the number of faces (*F*) vertices (*V*) and edges (*E*) for each regular polyhedron as well as the square-based pyramid and the hexagonal prism. Then complete the chart below.

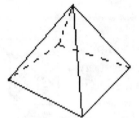   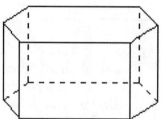

	Faces (F)	Vertices (V)	Edges (E)	Shape of Faces
Regular Tetrahedron				
Regular Hexahedron				
Regular Dodecahedron				
Regular Octahedron				
Regular Icosahedron				
Square-based Pyramid				
Hexagonal Prism				

B. Write a formula relating the numbers *F*, *V*, and *E*. This relationship is known as *Euler's Formula*.

C. Following the instructions on Appendix Pages 19 and 20, make a stellated octahedron (referred to as a *stella octangula*, Visual Geometry Project, Key Curriculum Press). Count the faces, vertices, and edges. Do the values for *F*, *V*, and *E* satisfy Euler's Formula?

D. View the phenomenal videotapes from Key Curriculum Press: *The Platonic Solids* and *Stella Octangula*.

Activity 11-16: Problem Solving with Pattern Blocks

Overview: Pattern blocks are used for investigating patterning/sequencing, for identifying special geometric shapes, for developing concepts of fractions (Activity 6-6) and area (Activity 12-7), and for exploring angle measures (Activity 12-6). This versatile manipulative is also used to develop spatial problem-solving skills and to make connections to numerical computations.

Materials: Pattern blocks (commercial or cut with Ellison Letter Machine)

A. Cover the following shape with pattern blocks. Record the number of each type of pattern block you used in row (7) of the table on the next page.

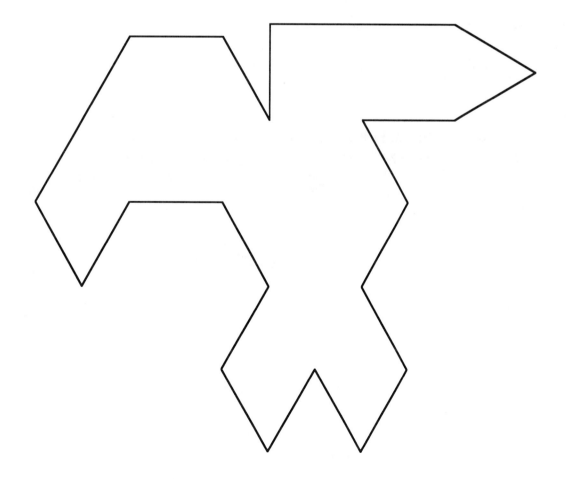

B. Cover the shape in Part A with different blocks to help you answer the questions below, and record your solutions in the table.

Covering	⬡	⬭	□	△	◇	⬭	"Value" of Covering
(1)							
(2)							
(3)							
(4)							
(5)							
(6)							
(7)							

(1) What is the fewest number of pieces that can be used in the covering?

(2) What is the greatest number of pieces that can be used in the covering?

(3) Can you cover the shape using pattern blocks of only one color?

(4) Can you cover the shape using pattern blocks of exactly two colors?

(5) Can you cover the shape using at least one block of each color?

(6) Can you cover the shape without using the tan rhombus?

C. Using the assigned values for each pattern block that you find below, determine the "value" of each covering in Part B and record in the last column of the table above.
Yellow hexagon = 6; red trapezoid = 3; orange square = 10; green equilateral triangle = 1; blue rhombus = 2; tan rhombus = 5

What do you observe?

Can you explain why?

Activity 11-17: Geometry in the World Around Us

Overview: Too often students "see" geometry only in the pages of a book. This activity/project requires that students find examples of geometry in the world around them and connect those examples to definitions and the traditional geometry drawings. In my classes, from elementary grades to graduate courses for teachers, we integrate a Geometric Tour through Historic Savannah with our projects. Anyone completing this project will "see" geometry in the world around us with different eyes.

Materials: Definitions of the geometry terms listed (See Appendix Pages 29 and 30); pictures of geometric objects from "real life," three ring binder. Note: As many assessments now include portfolio entries, this activity is suitable for such an assessment.

A. Become comfortable with definitions for the following terms:

   First Set: angle, acute angle, right angle, obtuse angle, zero angle, straight angle, adjacent angles, vertical angles, complementary angles, supplementary angles, simple closed curve, polygon, convex polygon, nonconvex polygon, triangle, acute triangle, right triangle, obtuse triangle, isosceles triangle, equilateral triangle, scalene triangle, parallel lines, skew lines, perpendicular lines

   Second Set: quadrilateral, parallelogram, rectangle, square, rhombus, kite, trapezoid, regular polygon, polyhedron, sphere, cube, prism

B. Choose a central theme or topic that will appeal to upper elementary/middle school students (Geometry Along the Road, Geometry in Architecture, Geometry in Historic Savannah, Geometry in my Neighborhood, Geometry in the Mall). The first set of definitions MUST relate to your central theme. You may not be able to fit some of those from the second set into your theme; just find examples -- do your best. Most students seem to be able to relate all terms to a chosen theme.

C. Prepare a notebook with a page for EACH of the geometric terms that includes THREE things: (1) the term and definition of the term at the top of the page, (2) a drawing of the geometric term (you may use templates or computer drawings) with only appropriate markings (no words or letters or numerals), and (3) a picture from a magazine or other source with the geometric figure correctly identified.

Notes:
*   The geometric figure in the picture should not be "too small."
*   Your markings should be obvious and *carefully* made on each picture.
*   Your pictures should be "real-life" examples; the geometric entity should be found **within** the picture, not made by the borders, etc.
*   Submit your handiwork in a three-ring binder with pictures glued on each page. (Some students like to put them in plastic "sleeves" before assembling the notebook.)
*   Your notebook should have a cover sheet that indicates the theme or the topic, and (of course) your name.

Activity 11-18: Motivating "Inclusive" Definitions

Overview: In the early grades, the first shapes children learn to identify are circles, triangles, and squares. Generally, after the square is introduced and discussed, the general rectangle is shown with the question: Is this shape a square? The answer, of course, is "no" and often there is a comment such as: "See, it has two longer edges and two shorter edges." Herein the difficulty lies. Children mistakenly conclude that a rectangle MUST have two longer edges and two shorter edges. This activity deals with the inclusion of special shapes (triangles, quadrilaterals) into more general "shape families" as a means of classifying them and seeing how they are related.

Materials: Several sheets of $8\frac{1}{2}$" x 14" paper to make an isosceles triangle and the general shapes: rhombus, trapezoid, kite, parallelogram.

A. If a polygon has four edges and four right angles, what is it called?_____
We consider a shape and ask the question: " Is this shape a rectangle?" Another way of determining whether or not a shape is a rectangle is to ask the two questions related to the definition: "Does the shape have four edges? Does the shape have four right angles?"

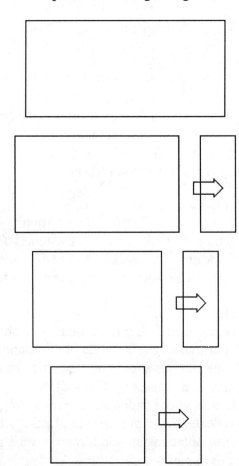

Take one sheet of the rectangular paper with two obviously longer edges. Ask the question: "Is this shape a rectangle?" OR: "Are there four edges? Are there four right angles?" The edges can be counted and each corner should be compared to a known square corner such as an index card.

Trim a rectangular strip from one end as shown and move it away. " Is the remaining shape a rectangle?" To determine the answer, you should verify there are four edges and four right angles.

Repeat the process, once again asking: "Are there four edges and four right angles?" Eventually, you will come to a place where the rectangle formed has an added feature: all edges have the same length.

We call this rectangle with congruent edges a square. So, a square is a special rectangle. As each shape that remains after the trimming fits the definition of rectangle, it is included in the set of rectangles. Using this reasoning, a square is included in the set of rectangles.

TEACHING NOTE:
In this text, we use the term "edge" instead of "side" in polygons. Note that when we speak of polyhedra, we use the terms "faces," "edges," and "vertices." Each edge is the intersection of two polygonal regions; that is, this segment is an edge of each polygon that is the boundary of a face.

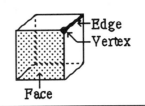

B. Cut an isosceles triangle whose height is about twice the base. Show it to your partner and ask: "Is this triangle isosceles?" In other words, when we consider the definition we will ask: "Does the triangle have two congruent edges?" Now make a symmetry fold. When one of the edges coincides with the other, one can readily see the two edges are congruent and the triangle is, therefore, isosceles.

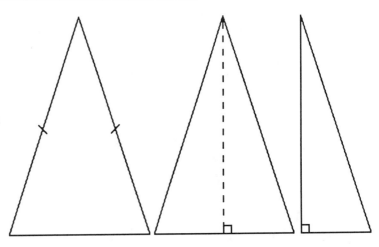

Trim a piece from the base vertices to a point on the symmetry fold as shown. Unfold the triangle and ask: "Does it have two congruent edges?" If so, then the triangle is isosceles. Of course, as the two oblique edges "exactly match" as seen when the triangle is folded, they are congruent. Refold and repeat the process.

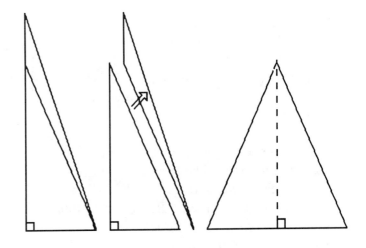

If we are careful, we will eventually make a trim so that not only will the oblique edges be congruent, but they will have the same length as the base. Again, the question is asked: "Does the triangle have two congruent edges?" Of course! Thus, this triangle, which we call equilateral, is an isosceles triangle and is included in the set of isosceles triangles.

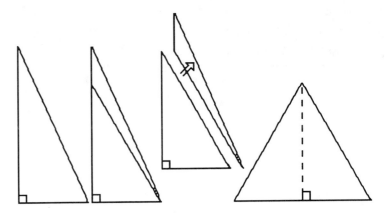

> **TEACHING NOTE:**
> As the symmetry fold in the isosceles triangle is perpendicular to and bisects the base, any point on this fold (a part of the perpendicular bisector) is equidistant from the endpoints of the base.

C. Cut a rhombus with acute and obtuse angles similar to the one shown. To determine if a shape is a rhombus, we must show that all four edges are congruent. Fold the rhombus in half along a vertical symmetry fold. Fold this triangular region in half along a horizontal symmetry fold. What is the shape formed by folding twice? _____ Notice that each edge of the rhombus "exactly matches" the others. Trim a portion from the vertex of the shorter leg of the right triangle to the other leg. Note that we have four "layered" right triangles with hypotenuses that "exactly match" and are congruent. Thus, when we unfold the shape, we have another rhombus.

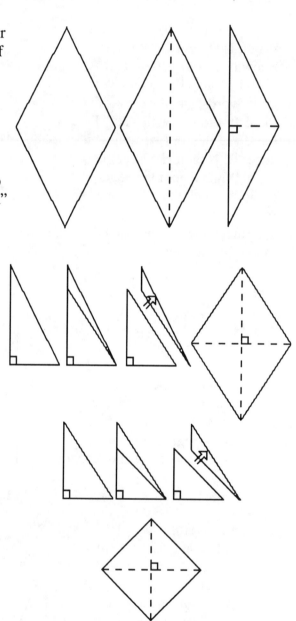

Repeat the process: fold twice and trim a portion by cutting from the vertex of the shorter leg to the other leg. Note, again, that the hypotenuses are congruent. When the shape is unfolded, once again we have a rhombus.

If we are careful, we can make a trim so that the resulting right triangle will be isosceles. When the shape is unfolded, we know it will once again be a rhombus. But, examine this shape closely. Get your square corner referent and compare it to the angles of this rhombus. The angles are square corners (right angles), so the rhombus now formed is a square. Thus, a square is included in the set of rhombuses.

TEACHING NOTE:
The same thing can be done to show that a rhombus is special kite. A kite is a quadrilateral with two distinct pairs of consecutive edges that are congruent. By making a symmetry fold and trimming as shown, we can see that a rhombus is included in the set of kites.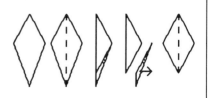

D. Cut a general parallelogram and, similar to the trimming of the rectangle to form a square, cut appropriate strips to show that a rhombus is included in the set of parallelograms. Discuss with a partner how the strips must be cut.

E. By anyone's definition, a trapezoid must have a pair of opposite edges that are parallel. From a long rectangular sheet of paper, cut off two right triangles from either end to obtain a trapezoid such as the one shown. Notice that the two horizontal edges are parallel. Trim and remove a triangular region by making a cut from the top right vertex to the bottom edge of the trapezoid. Consider the remaining quadrilateral and ask: "Is there a pair of parallel edges?" Thus, the new shape is also a trapezoid.

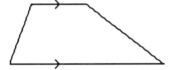 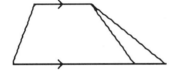 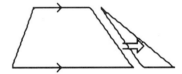

Repeat the process. Note that if we are careful, we can make a cut so that the two base angles are congruent and the two slant edges are congruent. What type of trapezoid is this?

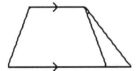 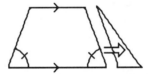

Continue the process. Again, if we are careful we can make a cut so that the resulting quadrilateral has the following properties: the two horizontal edges are both parallel and congruent and the two oblique edges are also congruent. Again, the question is asked: "Is there a pair of parallel edges?" Of course! So this shape, once again, is a trapezoid. However, the two oblique edges are also parallel. What is the name of a quadrilateral with both pairs of edges parallel? _____

Thus, the trapezoid now formed is a parallelogram, and we can see that a parallelogram is included in the set of trapezoids.

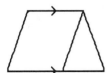 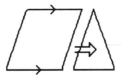

F. From a rectangular sheet of paper, cut a trapezoid with two square corners (a trapezoid having an edge perpendicular to the bases). Trim off triangular regions from the obtuse vertex to the opposite edge until a trapezoid is formed in which the two bottom base angles are congruent and the two vertical edges are congruent. What special name is given to this shape? _____ Justify that it is included in the set of isosceles trapezoids.

TEACHING NOTE:
Determining how to cut these general shapes from $8\frac{1}{2}$" × 14" rectangular paper is a problem-solving activity in itself. Discussing with others the various characteristics of the figures that help us decide how to cut them is a valuable experience.

TEACHING NOTE:
These concrete activities can be shown dynamically using software such as *Geometer's Sketchpad®*.

Below is a network of connections among seven types of quadrilaterals. It is quite useful because it allows us to relate the properties of specific quadrilaterals. *Any property true of all figures of one type in the hierarchy is also true of all figures of all the types below it to which the first type is connected.*

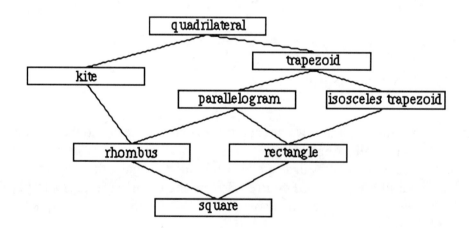

If you always do what you always did, you'll always get what you've always got.

Woodrow Wilson Algebra Team IV

Activity 11-19: Making a Mark

Overview: Younger students may identify a figure as a square because "it looks like a square." Mathematicians communicate about the attributes of geometric entities by marking them in special ways. It is important for students to "read" markings on geometric drawings so they may correctly identify the figures. For example, a square corner marking is used to denote a right angle; arrow markings are used to indicate parallelism; similar tick marks are used to indicate congruence just as using different numbers of tick marks indicates that two segments are not congruent. Students must be cautioned not to assume characteristics of a shape based solely on "how it looks."

Materials: Appendix Pages 29 and 30 for reference.

Directions: Work in pairs to determine the most specific name for each geometric figure based on the information given by the markings. Recognize that some markings call your attention to specific properties. Then discuss what additional names may be given to the figures if other markings were present. For example, in the first shape you can tell that all 10 edges are congruent, so the shape is an equilateral decagon. If a segment joining nonconsecutive vertices were drawn, you would say it is a nonconvex equilateral decagon. In the second shape, because there is a square corner along with parallel markings, you know the shape is a rectangle. Without the square corner, you could only say the shape is a parallelogram.

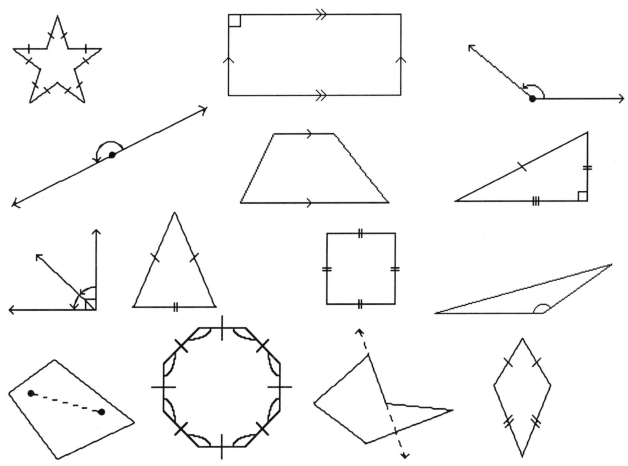

TEACHING NOTES: A SUMMARY

Geometry must be emphasized throughout the elementary and middle grades and should be infused into every area of the mathematics curriculum.

In emphasizing geometry there are some issues to be addressed. Students should be given opportunities to develop visualization skills. One tool that provides these opportunities is the geoboard. In using geoboards, students may be given a list of geometric figures (square, circle, right triangle, equilateral triangle, oval, etc.) and asked to form them. Then students would be asked to determine which figures can be and cannot be made on the geoboard and discuss reasons why they may be successful or not. They would probably observe that only figures with straight line segments can be made or that no equilateral triangles can be made.

Emphasis should be placed on the development of concepts from the very beginning. In the early grades, the identification of circles, triangles, and squares usually precedes rectangles. Once students can identify squares, the teacher may ask them to look at a non-square rectangular shape cut out of paper and determine whether or not it is a square. This is often accompanied by statements such as: *This is not a square; it is a rectangle. See, it has two longer edges and two shorter edges.* Children often conclude, after such examples, that a square is not a rectangle. This common misconception can be avoided if the rectangle is introduced (prior to the square) as a figure with four edges and four square corners. The teacher can then take a general rectangular shape from colored paper, cut off a strip, and ask the children whether or not the new shape is a rectangle. The students identify the shape as a rectangle, as it has four edges and four square corners. The teacher should continue to cut strips, leaving rectangular shapes, until all four edges of the "new" rectangular shape are the same length. Children identify the shape as having four edges and four square corners -- a rectangle. The teacher should point out that this shape is a special rectangle, called a square, since all four edges have the same length. Thus, the concept of a square as being a special rectangle having congruent edges is developed. Questions like *"Is this shape a square or is it a rectangle?"* should be avoided unless students are expected to answer: *"It is both!"* As these and other concepts of geometric shapes are being developed, teachers should be sure to give both examples and nonexamples of concepts discussed.

Students need to eventually differentiate between *circle* and *circular region* or between *square* and *square region*. Younger children may be directed, on work sheets, to color the circles green, the triangles blue, and so on. Then they color inside the shapes. This often leads to their perception of the circular region as the "circle" or the triangular region as the "triangle." To circumvent this misconception, directions such as *"Trace the figures that are circles with a green crayon"* or *"Color the inside of the rectangles which are not squares with a red crayon"* should be given instead.

A unit in geometry is a wonderful place to ask students to be creative in writing activities. Here is a poem written by a seventh grader (after studying the classification of polygons):

OCT-
Octave: eight tones.
Octet: eight voices.
Octane: eight carbon atoms.
Octameter: eight metrical feet.
Octapeptide: eight amino acids.
Octopus: eight tentacles.
Octagon: eight angles.
Octogenarian: eight decades old.
Octosyllabic: eight syllables.
October: **Why not the eighth month?**
Mandy, 1990

Consider the geometry standard from the NCTM *Principles and Standards for School Mathematics*:

GEOMETRY STANDARD

Instructional programs from prekindergarten through grade 12 should enable all students to—

- *Analyze characteristics and properties of two- and three-dimensional geometric shapes and develop mathematical arguments about geometric relationships*

- *Specify locations and describe spatial relationships using coordinate geometry and other representational systems*

- *Apply transformations and use symmetry to analyze mathematical situations*

- *Use visualization, spatial reasoning, and geometric modeling to solve problems*

RESOURCES

Periodicals:

Andrews, Angela Giglio. "Developing Spatial Sense-a Moving Experience." *Teaching Children Mathematics* 2 (January 1996): 290-93.

Bartels, Bobbye Hoffman. "Truss(t)ing Triangles." *Mathematics Teaching in the Middle School* (Mar.-Apr. 1998): 156-61.

Battista, Michael T. " Geometry Results from the Third International Mathematics and Science Study." *Teaching Children Mathematics* 5 (February 1999): 367-373.

Carroll, William M. "Middle School Students' Reasoning about Geometric Situations." *Mathematics Teaching in the Middle School* 3 (Mar. - Apr. 1998): 398-403.

Chappell, Michaele F. " Geometry in the Middle Grades: From Its Past to the Present." *Mathematics Teaching in the Middle School* 6 (May 2001): 516-519.

Clements, Douglas. " Young Children's Ideas about Geometric Shapes." *Teaching Children Mathematics* 6 (April 2000): 482-488.

Coes, Loring. "Building Fractal Models with Manipulators." *Mathematics Teacher* 86 (November 1993): 646-651.

Dunkels, Andrejs. "Making and Exploring Tangrams." *Arithmetic Teacher* 37 (February 1990): 38-42.

Emenaker, Charles E. " Gingerbread-House Geometry." *Teaching Children Mathematics* 6 (December 1999): 208-215.

Fuys, David J., and Amy K. Liebov. "Concept Learning in Geometry." *Teaching Children Mathematics* 3 (January 1997): 248-51.

"Geometry Must Be Vital." *Teaching Children Mathematics* 7 (March 2001): 409-415.

"Geometry Projects Linking Mathematics, Literacy, Art, and Technology." *Mathematics Teaching in the Middle School* 4 (February 1999): 332-335.

"Geometry Through Beadwork Designs." *Teaching Children Mathematics* 7 (February 2001): 362-367.

Glidden, Peter L., and Erin K. Fry. "Illustrating Mathematical Connections: Two Proofs That Only Five Regular Polyhedra Exist." *Mathematics Teacher* 86 (November 1993): 657-661.

Hannibal, Mary Anne. "Young Children's Developing Understanding of Geometric Shapes." *Teaching Children Mathematics* 5 (February 1999): 353-357.

Hopley, Ronald B. "Nested Platonic Solids: A Class Project in Solid Geometry." *Mathematics Teacher* 87 (May 1994): 312-18.

Izard, John. "Developing Spatial Skills with Three-Dimensional Puzzles." *Arithmetic Teacher* 37 (February 1990): 44-51.

Onslow, Barry. "Pentominoes Revisited." *Arithmetic Teacher* 37 (May 1990): 5-9.

Pearson, Jennifer, and John Wallace. "What Makes a Corner a Corner?" *Teaching Children Mathematics* 5 (September 1998): 6-9.

Pereira-Mendoza, Lionel. "What Is a Quadrilateral?" *Mathematics Teacher* 86 (December 1993): 774-776.

"Picture This: Second Graders "See" Symmetry and Reflection." *Teaching Children Mathematics* 7 (December 2000): 208-209.

Rigdon, Deanna, Jolyn Raleigh, and Shari Goodman. "Math by the Month: Pattern-Block Explorations." *Teaching Children Mathematics* 6 (November 1999): 182-183.

Rigdon, Deanna, Jolyn Raleigh, and Shari Goodman. "Math by the Month: Tackling Tangrams." *Teaching Children Mathematics* 6 (January 2000): 304-305.

Robertson, Stuart P. "Getting Students Actively Involved in Geometry." *Teaching Children Mathematics* 5 (May 1999): 526-529.

Rubenstein, Rheta N., Glenda Lappan, Elizabeth Phillips, and William Fitzgerald. "Angle Sense: A Valuable Connector." *Arithmetic Teacher* 40 (February 1993): 352-358.

"Shaping the Standards: Geometry and Geometric Thinking." *Mathematics Teaching in the Middle School* 4 (February 1999): 301.

Swindal, Donna Norton. "Learning Geometry *and* a New Language." *Teaching Children Mathematics* 7 (December 2000): 246-250.

Tepper, Anita Benna. "A Journey through Geometry: Designing a City Park." *Teaching Children Mathematics* 5 (February 1999): 348-352.

Thatcher, Debra H. "The Tangram Conundrum." *Mathematics Teaching in the Middle School* 6 (March 2001): 394-399.

"The Angles of a Star." *Mathematics Teacher* 93 (September 2000): 512-516.

"The Earliest Geometry." *Teaching Children Mathematics* 7 (October 2000): 82-86.

van Hiele, Pierre M. "Developing Geometric Thinking through Activities That Begin with Play." *Teaching Children Mathematics* 5 (February 1999): 310-316.

Wilson, Patricia, and Verna Adams. "A Dynamic Way to Teach Angle and Angle Measurement." *Arithmetic Teacher* 39 (January 1992): 6-13.

Woodward, Ernest, and Rebecca Brown. "Polydrons and Three-Dimensional Geometry." *Arithmetic Teacher* 41 (April 1994): 451-58

Books and Other Literature:

Charles, Randall I., D. A. Seldomridge, G.A. Seldomridge, and Robert P. Mason. *Problem-Solving Experiences in Geometry* (1991). Addison-Wesley Publishing Company.

Foster, Thomas E. *Tangram Patterns* (1977). Creative Publications.

Goodnow, Judy, S. Hoogeboom, and A. Roper. *Moving On with Tangrams: Intermediate Problem-Solving Activities* (1988). Creative Publications.

Lindquist, Mary J., and Albert P. Shulte. *Learning and Teaching Geometry, K-12* (1987). The National Council of Teachers of Mathematics, Inc.

Paul, Ann Whitford. *Eight Hands Round: A Patchwork Alphabet.* New York: HarperCollins Publishers, 1991.

Pedersen, Katherine, et al. *Trivia Math-Geometry & Advanced Topics* (1987). Creative Publications.

Picciotto, Henri. *Pentomino Activities, Lessons, and Puzzles* (1984). Creative Publications.

Picciotto, Henri. *SuperTangrams for Beginners* (1987). Creative Publications.

Pugh, Anthony. *Polyhedra: A Visual Approach* (1976, 1990). Dale Seymour Publications.

Seymour, Dale. *Tangramath* (1976). Dale Seymour Publications.

Usiskin, Zalman, et al. *The University of Chicago School Mathematics Project GEOMETRY* (1997). ScottForesman Publishing Company.

Other Resources:

Brest, Hilary, Ann E. Fetter, Cynthia Schmalzried, Doris Schattschneider, and Eugene Klotz. *Stella Octangula* (1991). Key Curriculum Press. *Activity Book.*

Eckert, Nancy, Ann E. Fetter, Cynthia Schmalzried, Doris Schattschneider, and Eugene Klotz. *Platonic Solids* (1991). Key Curriculum Press. *Activity Book.*

Key Curriculum Staff. *Platonic Solids* (1991). Key Curriculum Press. *Videos and Manipulative Kit.*

Key Curriculum Staff. *Stella Octangula* (1991). Key Curriculum Press. *Videos and Manipulative Kit.*

Moore, Peter. *POLY-CUBE-IT©.* Blue Marble Bookstore. *Manipulative Kit.*

Polydron™. Cuisenaire Company of America, Inc. *Shapes and Protractor.*

12

MEASUREMENT

> *Physical action is one of the bases of learning. To learn effectively the child must be a participant in events, not merely a spectator. To develop his concepts of number and space, it is not enough that he look at things. He must also touch things, move them, turn them, put them together, and take them apart. For every new concept that we want the child to acquire, we should start with some relevant action that he can perform. - Irving Adler*

What does it mean "to measure"? For any given attribute (length, area, volume, temperature, mass/weight, angle), the process of measuring is the same. Pick an attribute to measure; select a unit to measure that attribute; compare that unit to the object being measured; and report the number and type of unit used. The number and type of unit can be determined in several ways; *e.g.*, counting, using a formula, or using a measuring device. Prior to using a formula or a measuring instrument, however, students should have ample opportunity to experience a variety of activities that allow them to compare objects directly, to "cover and count" (area is a measure of covering) or to "fill and count" (volume is a measure of filling). If formulas or instruments are used too soon, students are left without the necessary understanding for solving measurement problems.

The first step in the measurement process cannot be taken for granted. If you were asked to measure a computer, what would you do? Would you measure its weight? Would you give the dimensions of the monitor? Would you measure the memory on the hard drive? Would you give its value (old, new, costly, cheap)? This first step of identifying the attribute to be measured is important. With some objects, the attribute being measured may be obvious; but, with others, it is not. The selection of a unit with which to measure depends on the context in which the measurement occurs. Any geometric shape which tessellates the plane can be used to measure area. Any one of the Platonic solids (which tessellate space) will serve as a unit for measuring volume. By understanding how to use nonstandard units and some of the difficulties with their use in communication, students will better appreciate the need for standard units.

Measurement is an area of the curriculum which allows students to see that mathematics is useful in everyday life. It is an excellent arena for applying concepts of fractions and decimals. Students should be involved in estimation as it aids in their understanding of the attributes of an object, the process of measuring, and the sizes and usefulness of various units. Just as an athlete gets better through practice, so students get better at estimation by practicing.

Some measures are discrete (the number of children in a room, the number of M&M's® in a jar, the number of days in a week) and some are continuous (how long it takes to get to school, the length of a finger, the degree measure of an angle, the area of an egg-shaped region). When reporting measurements, students should give the measure by designating both a *count* (a quantity) and a *counting unit* (the unit of measure); *e.g., 32 children, 5 inches, 11.5 grams, 3 cm²*. As many measurements are not exact, students should recognize that it is often appropriate or necessary to report a measurement as "about 3 minutes" or "between 44 and 45 centimeters."

Activity 12-1: Meter Measure

Overview: Students can better understand how to use a measuring device if they participate in its construction. Estimating lengths before actually measuring is a critical part of developing a "feel" for units of length. In this activity you will make a "ruler" which is 1 meter in length. Each segment of the ruler is 1/10 of a meter (or 0.1 meter), which is a decimeter. One of the decimeter segments is divided into 10 parts, so that measurements can be made to the nearer centimeter.

Materials: 2 cm × 10 cm paper strips of assorted colors (easily cut with Ellison Letter Machine™ and the Fraction Square $\frac{1}{5}$ die); 1 cm × 1 cm squares in assorted colors (easily cut with Ellison Letter Machine™) or a 1 cm × 10 cm strip cut from centimeter grid paper (Appendix Page 26)

A. Tape 10 of the 2 cm × 10 cm strips together end-to-end so that no two adjacent strips have the same color. The length of these strips together is one meter. Find some length on your body that is 1 meter in length (such as the distance from your fingertips on your outstretched arm to your nose).

B. Using your meter "ruler" for measuring, find (to the nearer meter) the following lengths in your classroom after you first estimate the lengths.

Object to be Measured	Estimate	Measurement
Length of your teacher's desk		
Distance from the floor to the doorknob		
Width of the overhead screen or chalkboard		
Length of the classroom		
Width of the classroom		
Height of your teacher (or a partner)		

C. Each of the ten strips you used to make your meter ruler is 1/10 of the meter and is called a decimeter. Estimate and then measure each of the following to the nearer decimeter (using your meter ruler).

Object to be measured	Estimate	Measurement
Head size		
Arm length		
Length of little finger		
Wrist size		
Length of shoe		
Your height		

D. Glue the 1 cm × 10 cm strip of grid paper or 10 of the centimeter squares edge-to-edge along the first decimeter length of your metric ruler.

Now you can measure objects to the nearer centimeter. Estimate and then measure each of the following lengths to the nearer centimeter, and complete the table below.

Object to be measured	Estimate	Measurement
Length of little finger		
Thickness of this activity manual		
Wrist size		
Length of pencil		
Width of this page		

E. Each centimeter length can be divided into ten equal parts of 1 millimeter in length. (A dime is about 1 mm thick.) Find two other objects that have measures of approximately one millimeter.

TEACHING NOTE:

When children first begin measuring the length of objects, they should use nonstandard units such as popsicle sticks, pencils, coffee stirrers, or paper clips. It would even be a good idea to use base ten rods as a nonstandard unit; later, students will learn that a rod is metricized and is a decimeter in length. When students begin to measure length in inches, they can lay one-inch squares end-to-end along an object and measure the object by counting the one-inch lengths. A teacher may ask students to find the width of the classroom door only to have children comment that the individual one-inch squares won't stay up on the door. The wise teacher will already have 1 in. × 12 in. strips cut from 9 × 12 construction paper on which the children can glue the multi-colored paper one-inch squares edge-to-edge to make a ruler. This ruler, of course, is the "foot" used in the U.S. Customary system. [Students often ask why we don't just use 10 of the square inches to make the ruler instead of twelve; grouping by tens is a natural extension of our numeration system. The metric system, based on groups of ten, is really easier for children to master.] Since the paper strip ruler is bendable (unlike commercial plastic or wooden rulers), students can measure their wrist sizes or the circumferences of their heads. They will better understand how to use a ruler if they are involved in constructing one. Additionally, students could fold a paper one-inch square in half and mark off half-inch markings on their rulers to then begin measuring to the nearer half inch. A similar process can be done with quarter-inch markings.

Activity 12-2: The Distance Around

Overview: Our intuition about the distance around a circle (the circumference) is often not very good. In this activity we will sharpen our skills of estimation and get a clearer understanding of the distance around the circle.

This activity should be completed by the class as a group.

Materials: Scissors, a trash can with a circular opening (circular top), a blue ball of yarn and a red ball of yarn (any two colors of yarn will do), and 8 - 10 drinking glasses of assorted sizes

A. Each member of the class should look carefully at the opening (the top) of the trash can and cut off a piece of blue yarn that represents his or her estimate of the circumference of that top. Each student should measure his or her piece to the nearer centimeter. The pieces of yarn representing the estimates of the class should be taped to the top of the door frame of the classroom in ascending order by length. Measure the actual circumference with a piece of red yarn, cut it, measure it to the nearer centimeter, and tape it in its place in the doorway.

What is the average of the estimated lengths? _____

What is the measure of the actual circumference? _____

Discuss the results. How does the actual measurement compare with the average of the estimates? Why do you think that this is true?

B. Select one of the glasses or cups at a time. Guess whether the glass is bigger around at the top than it is tall, smaller around the top than tall, or the same length around the top as it is tall. Take a piece of yarn and measure around the top of the glass. How do your initial guesses compare to your later guesses? Can you identify some characteristic(s) of the glasses which are taller than they are round?

TEACHING NOTE:
This activity involving guessing whether a glass is bigger around than it is tall allows students to make a conjecture, measure to compare, and then adjust perceptions of "roundness" in comparison to "tallness". The processes of estimate/conjecture, compare, reflect, and adjust are important ones in measurement.

Activity 12-3: Weighing In

Overview: Even though technically not the same, students may consider weight and mass as equivalent measures in everyday use. Weight is the force on an object due to gravitational pull. Mass is the amount of matter of an object. For example, you would weigh less on the moon than on earth, as the gravity on the moon is less than on earth. However, your mass is the same on earth as on the moon. Once again, before using standard units, students should first compare weights of objects to order them from lightest to heaviest. Then students should use nonstandard units to weigh objects.

Materials: Balance scale; Unifix® cubes, wooden blocks, Hershey Kisses®, or base ten rods; a one-kilogram mass; a one-gram mass; familiar objects from a classroom or from home.

A. Choose six objects (*e.g.*, chalkboard eraser, pencil, notebook, box of crayons, scissors, bottle of glue) and order them from lightest to heaviest by picking up the objects and comparing them. Check your ordering using a balance scale (but do not use any units of measure).

B. Choose a unit of mass/weight like Hershey Kisses®, wooden blocks, or base ten rods. First, estimate the weight of each of the six objects in your chosen unit. Use the balance scale to find the actual measurement and record your data below.

Object to measure in chosen unit	Estimate	Measurement

C. Hold a 1 kilogram mass as a referent. After experiencing how heavy it feels, select 6 - 8 objects in your classroom to use for comparison to 1 kilogram. Hold one object at a time to determine whether it weighs more than, less than, or the same as 1 kilogram. Check your conjecture using a balance scale.

TEACHING NOTE:
When introducing children to metric weight measurement, the kilogram should be used first. The gram, even though it is the basic unit of metric weight, is too small for children to use conveniently in estimating weight.

D. Hold a one-gram mass in your hand to experience the "feel" of one gram. Choose some light objects to compare to the gram to determine which are lighter than, heavier than, or the same mass as one gram. Complete the following table.

Object to be measured	Estimate	Measurement
1 paper clip		
1 M&M's® candy		
a dime		
a pencil eraser		
a thumb tack		

Activity 12-4: Estimilliliteration

Overview: This activity allows students to estimate and measure in milliliters in a game-like format.
This game is played in groups of four persons.

Materials: Graduated cylinder marked in milliliters up to 100 ml, assortment of cups or other containers which do not show capacity markings, and a source of water

One student is the judge and declares a certain measure less than 100 ml. The other students each record this amount on their charts. Each will take an unmarked container and pour into it his or her estimate of the declared amount. In turn, each student gives the estimated amount to the judge, who checks it by pouring the water into the graduated cylinder and reading the amount. The actual measured amount is recorded on the chart and the size of the estimation error is determined by subtracting. The student with the smallest error receives one point. The first student with 5 points wins. To allow each student to compete, group members should take turns being judge.

Amount declared by judge	Amount I gathered	How far off was I?
40 ml	83 ml	43 ml
__ ml	__ ml	__ ml
__ ml	__ ml	__ ml
__ ml	__ ml	__ ml
__ ml	__ ml	__ ml
__ ml	__ ml	__ ml
__ ml	__ ml	__ ml
__ ml	__ ml	__ ml
__ ml	__ ml	__ ml
__ ml	__ ml	__ ml
__ ml	__ ml	__ ml
__ ml	__ ml	__ ml
__ ml	__ ml	__ ml
__ ml	__ ml	__ ml
__ ml	__ ml	__ ml

Activity 12-5 : Is a Cup Really 8 Ounces?

Overview: In the U.S. Customary System there is some ambiguity in the unit of measure "ounces." Students learn there are 8 ounces in a cup but there are 16 ounces in a pound. Depending on the context, ounces can be a liquid or dry measure (filling) or can be a measure of mass or weight.

Materials: A one-cup measuring cup; Ziploc® baggies; dry oatmeal, M&M's®, sand, rice, gravel, water; scale that will weigh in ounces; modeling clay or Playdoh®.

A. Take off an amount of modeling clay or Playdoh® that you think weighs one ounce. Use the scale to weigh it. Did you underestimate or overestimate or are you just right? Adjust your hunk of clay or Playdoh® so that it does indeed weigh one ounce. *Feel* the weight of one ounce. Get fifteen more clay hunks of one ounce each (get your classmates involved) and put the 16 ounces of clay together. *Feel* the weight of the 16 ounces or one pound.

B. Using the measuring cup, measure out one cup each of dry oatmeal, M&M's®, sand, rice, gravel, and water and seal each in a Ziploc® baggie. By holding the baggies and comparing, seriate from lightest to heaviest. Note that you have measured out one cup, or 8 ounces, of each of the substances.

C. Weigh your baggies, beginning with your estimate of the lightest and moving on to the heaviest. Which of the 1 cup (8 ounces by volume) amounts was less than 8 ounces by weight? was more than 8 ounces by weight? was 8 ounces by weight?

TEACHING NOTE:

When my daughter Cassie was in an early elementary class, she came home and replied to my asking what she had done that day in mathematics: "We learned about ozzes and libs." My first thought was something like "This must be similar to *3 glips in a glop and 4 glops in a gloop; how many glips in 2 gloops?*" When I inquired further, she said, "You know, Mom. There are 16 ozzes in a lib." It dawned on me that she meant ounces (oz = ozzes) and pounds (lb = libs). She could determine *How many ounces in 2 pounds?* and *How many ounces in 3 pounds?* but she was clueless as to what an ounce or a pound were. She asked: "Mom, what does an ounce look like?" I tried to explain (NOT!) but then got out the Weight Watcher's scale (never been used) and a hunk of cheese. I cut off what I estimated to be an ounce of cheese and weighed it. After making the necessary adjustment, I let her feel the weight of one ounce of cheese. After learning to read the scale, she spent the next half hour cutting off celery, counting out M&M's® and sticks of gum, and gathering little boxes or figurines around the house to weigh after she had guessed whether they weighed more or less than an ounce. Afterwards, she said, "Mom, please let me take this to school. I don't think my teacher knows about this." IT IS SO IMPORTANT THAT WE ALLOW CHILDREN TO TOUCH AND FEEL AND MEASURE AND EXPERIENCE!

Activity 12-6: Angles with Pattern Blocks

Overview: When students are typically introduced to the standard unit for measuring angles, the degree, they learn that there are 90° in a right angle or perhaps 180° in a straight angle or 360° in one revolution of a circle. Pattern blocks provide a nice way to find the degree measure of each angle while using a familiar manipulative. In the diagram to the right, the sum of the measures of the obtuse angles of three blue rhombuses is 360°. So, each obtuse angle has a measure of 120°.

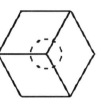

    Work with a partner.

Materials: Pattern blocks

A. Determine the measure of the angles in each of the Pattern Blocks below. Discuss with a partner how you determined each angle measure.

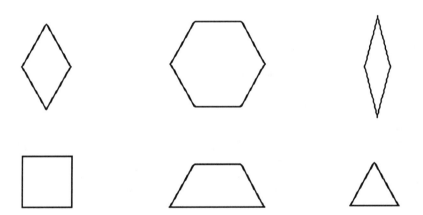

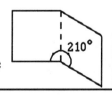

TEACHING NOTE:
Even though angles with measures greater than 180° are not usually introduced in the middle grades, they are certainly accessible when coupled with a manipulative like pattern blocks. For example, the nonconvex hexagon to the right has an interior angle whose measure is 210° (90° from the right angle of the square and 120° from the obtuse angle of the blue rhombus).

B. Cover the following shape with pattern blocks. Trace around each shape to record. Find the measure of each interior angle.

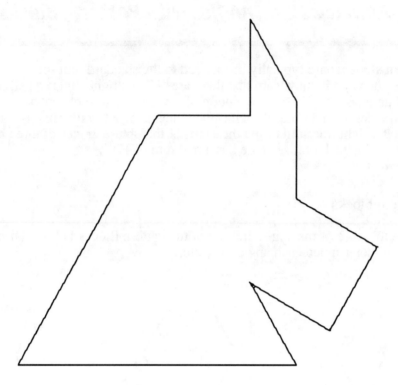

C. Get a partner. Take one each of the six pattern blocks. Make a shape by fitting the six blocks together. After tracing around the shape, determine the angle measure of each interior angle of the shape. Trade with your partner. Each student should then cover his/her partner's shape with pattern blocks and determine the angle measure of each angle. Compare results.

Activity 12-7: Developing the Concept of Area

Overview: The concept of area is "cover and count." Students' first experiences should involve covering planar regions with nonstandard units prior to using standard units. Any planar shape which "tiles" will suffice as a unit of area.

Materials: Pattern blocks: yellow hexagon, red trapezoid, blue rhombus; tangram pieces (Appendix Page 18), one-inch squares, and square dot paper (Appendix Page 21)

A. Estimate and then find the area of the shape using the following units of area: the yellow hexagon, the red trapezoid, and the blue rhombus.

Area = _____ hexagons

= _____ trapezoids

= _____ rhombuses

B. Assume that the area of the parallelogram from the tangram puzzle is 1. What is the area of this shape?

C. The following shapes are drawn on isometric dot arrays. The basic unit of area is the equilateral triangle.

    (a) Find the area of shape *A*.

        The area is ___ triangles.

    (b) By comparing shape B to shape A, find the area of shape *B*.

        The area is _____.

    (c) As shape B above has an area of one triangle, then it, too, can be used as a unit of area. Draw segments dot-to-dot to triangulate each region below to determine the area.

D. Estimate the area of each region in square inches. Then cover each shape with one-inch squares to check your estimate.

        Estimate: _____

        Actual Measure: _____

        Estimate: _____

        Actual Measure: _____

E. Square dot paper pictorially models work with the geoboard. The basic square unit of area is shown. Additionally, a half-square of area $\frac{1}{2}$ is shown. Divide each of the following regions into squares and half-squares to determine its area.

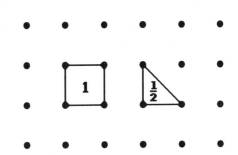

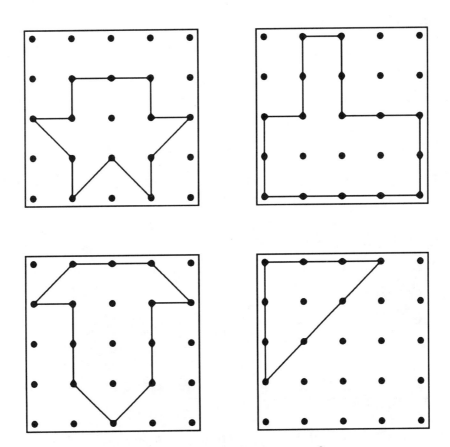

F. Which shapes in E have symmetry folds? Which are convex?

G. On square dot paper, make six different polygons with areas of 8 square units.

Activity 12-8: Area of Right Triangles from Rectangles

Overview: The rectangular array area model for multiplication was developed in Activity 4-10. Students should have discovered that the area of a rectangle, in square units, is the product of its two dimensions. Base and height are the names given to the adjacent edges of the rectangle. If the rectangle is divided in half along a diagonal, the two halves are congruent right triangles. The leg lengths of the right triangle are the dimensions of the related rectangle. Thus, the area of a right triangle is just *one-half of the area of its related rectangle.*

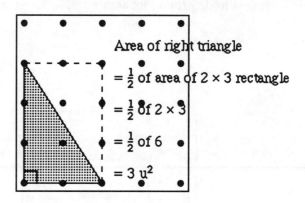

Area of right triangle
$= \frac{1}{2}$ of area of 2×3 rectangle
$= \frac{1}{2}$ of 2×3
$= \frac{1}{2}$ of 6
$= 3\ u^2$

Materials: 11×11 geoboard and square dot paper (Appendix Page 21)

A. Draw dotted segments to indicate the related rectangle for each right triangle. Label the dimensions. Find the area of the right triangle.

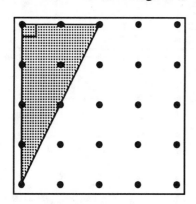   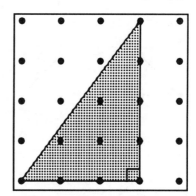

B. On an 11×11 geoboard, find all distinct right triangles having the given area. Draw solutions on square dot paper.

(a) $8u^2$

(b) $4\frac{1}{2}u^2$

Activity 12-9: Area–Rectangle Wrap Around

Overview: Determining the area of polygons on a geoboard or geodot paper is accessible to students who know two things: (1) the area of a rectangle is the product of its two dimensions and (2) the area of a right triangle is half the area of the related rectangle. For example, to find the area of polygon A, "wrap" the smallest possible rectangle around the polygon. From the area of the wrapped rectangle, subtract the areas of the two right triangles which lie outside polygon A, but inside the wrapped rectangle. This strategy is called "rectangle wrap around." It is simple, yet powerful. If you do the work on a geoboard, you truly will wrap the geoband around the polygon whose area you wish to find.

Work with a partner.

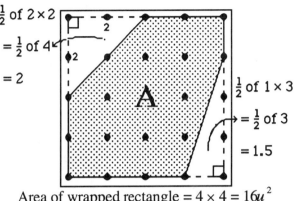

$\frac{1}{2}$ of 2×2

$= \frac{1}{2}$ of 4

$= 2$

$\frac{1}{2}$ of 1×3

$= \frac{1}{2}$ of 3

$= 1.5$

Area of wrapped rectangle $= 4 \times 4 = 16u^2$
Area of shaded region $= 16 - 2 - 1.5 = 12.5u^2$

Find the area of each shaded region by using the rectangle wrap around strategy. Discuss with a partner other strategies for finding the area. Are there any shapes below whose areas are difficult to find without using rectangle wrap around? [Observe that there are elementary strategies (other than rectangle wrap around) for finding area in (a), (b), and (d); but (c) is difficult unless more sophisticated "tools" are used (such as finding the exact length of the edges of the triangle and using Heron's Formula.]

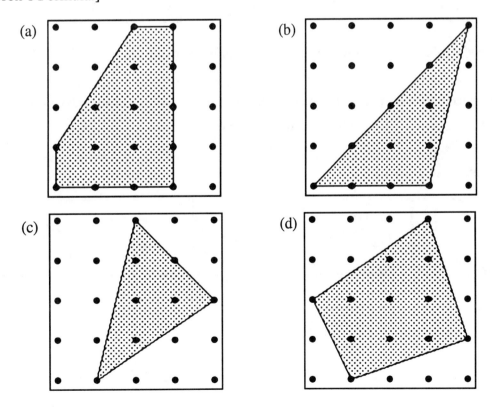

(a)

(b)

(c)

(d)

Activity 12-10: Area of Plane Figures

Overview: The area of a rectangular region is given by the product Area = (base)(height). [Children should feel comfortable with this fact since they should have used the rectangular array area model in studying the properties of multiplication (See Activity 4-10).] In this activity we will use paper cutting and taping to determine area formulas for parallelograms, triangles, and trapezoids using the fact that we know how to compute the area of a rectangle. Further, we will give a heuristic argument suggesting the formula for the area of a circle.
Work in groups of 3 or 4 persons.

Materials: Two congruent circles (two different colors) cut from construction paper, construction paper, scissors, and tape

A. **The Area of a Parallelogram:**

(a) Each member of the group should cut a non-rectangular parallelogram and label the base *b* and the height *h*. (Note that you can find the height *h* by folding the vertex A of the parallelogram back until it lies on the base of the parallelogram and then form a crease. The crease is perpendicular to the base; hence, its length is the height of the parallelogram.)

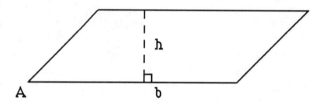

TEACHING NOTE:

If students are perplexed about how to cut a parallelogram from a rectangle, suggest these steps:

(1) Cut an oblique segment from one vertex to the opposite edge.

(2) Rotate the triangular shaped piece so obtained and slide it to the opposite edge of the rectangle.

(3) Cut off a congruent triangle from that end of the rectangle.

Additionally, making the parallelograms from square grid paper may make this process more understandable to some students.

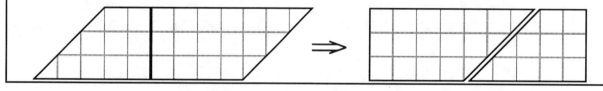

(b) Mark the right angles on either side of the crease and then cut along that crease. Slide one of the resulting parts horizontally to the end of the other part, matching slant edges together.

(c) The members of the group should now discuss the following questions about their parallelograms and the new figure formed from their parallelograms.

What is the new shape formed by the two pieces? _____

If the length of the base of the original parallelogram was b, what is the length of the base of the new shape? _____

If the height of the original parallelogram was h, what is the height of the new shape? ____

Because cutting and reassembling does not affect area measure, what is the area of the new shape and, thus, the parallelogram?

Area of parallelogram = _____

B. **Area of a Triangle:** Each member of the group should now cut a pair of congruent triangles. Care should be taken that one pair is obtuse and scalene, another should be right scalene, another acute isosceles, and so on.

(a) On each triangle clearly label a base and a corresponding height. The altitude can found by folding the base onto itself so that a crease is made which is perpendicular to the base and passes through the vertex opposite the base.

(b) Starting with one triangle placed on top of the other, rotate one so that a pair of corresponding edges match.

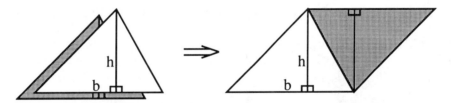

(c) Discuss in your group the answers to the following questions about the new figure you have formed.

The new figure is a _____. (Justify!)

If the length of the base of the triangles is b, the base of the new figure is _____.

If the height of the triangles is h, the height of the new figure is _____.

The area of the new figure is _____.

The area of *one* of the two congruent triangles is given by

Area = _____.

C. **Area of Trapezoids:** Each member of the group should cut a pair of congruent trapezoids from paper.

   (a) On each trapezoid label the two parallel edges b_1 and b_2. Fold to find the height and label it h on each trapezoid.
   (b) Starting with one trapezoid on top of the other, rotate one so that a pair of corresponding edges match.

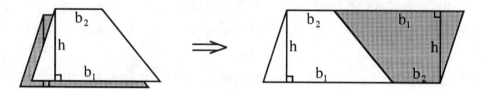

   (c) Discuss in your group the answers to the following questions about the new figure you have formed.

   The new figure is a _____.   (Justify!)

   The base of the new figure is _____.

   The height of the new figure is _____.

   The area of the new figure is _____.

   The area of *one* of the two congruent trapezoids is given by

   Area = _____.

D. **The Area of a Circle:** Each group should take two congruent circles, one of each of two different colors. Fold each circle in half and then cut along the fold. Each group should keep one semicircular region of each color and lay the other two semicircular regions aside.

   (a) Fold the two semicircular regions in half and cut again. Repeat the process until each semicircular region has been cut into 8 congruent regions (each piece is one sixteenth of the original circular regions).

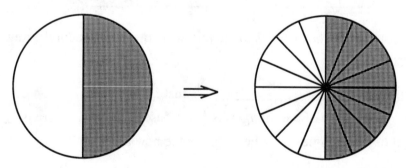

(b) Alternating colors, tape the sixteenths together forming a shape that looks somewhat like a parallelogram whose height is approximately r, the radius of the circular regions, and whose base is formed of short arcs. The sum of the lengths of those short arcs is half the circumference of the circular regions, or $\frac{1}{2}C = \frac{1}{2} \times 2\pi r = \pi r$.

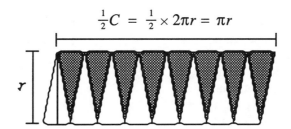

$$\tfrac{1}{2}C = \tfrac{1}{2} \times 2\pi r = \pi r$$

(c) Think of the "parallelogram-like" region as a parallelogram.

The base is approximately _____.

The height is approximately _____.

The area of the circle (cut and reassembled as the area of the "parallelogram-like" region) is then approximated by the formula

base × height = _____ × _____.

Observe that if the semicircular regions had been cut into more and more pieces, then the "arcs" making up the edges of the "parallelogram-like" regions would look even more like a straight line and the figure itself would look even more like a parallelogram. We reason that as the semicircular regions are cut into more and more pieces and are used to form "parallelogram-like" regions that the limiting value of the "parallelogram-like" regions is a parallelogram with height r and base πr. Thus, the area of the circle is given by the formula

Area = _____

Activity 12-11 Lengths on a Geoboard

Overview: The basic unit of length on a geoboard or on geodot paper is the horizontal or vertical distance between adjacent pegs or dots. Any oblique segment on the geoboard is the hypotenuse of a right triangle. Most of these oblique, or slanted, segments have irrational lengths. In this activity, you will use the Pythagorean Theorem, which is investigated in Activities 13-2 and 13-3.
Work with a partner.

Materials: Geoboard and geobands; pictorial geoboard

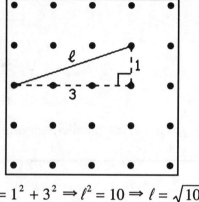

$$\ell^2 = 1^2 + 3^2 \Rightarrow \ell^2 = 10 \Rightarrow \ell = \sqrt{10}$$

A. Find all possible distinct lengths (give exact lengths) which can be made on a 5 × 5 geoboard. Organize and discuss your reasoning. Draw one each of the segments having different lengths on the pictorial geoboard below.

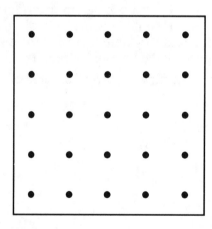

B. Seriate the different lengths from least to greatest.

C. What is the probability that a segment drawn at random will have a length greater than two?

Activity 12-12: Comparing Volumes

Overview: In general, our intuition about volumes is not very good. In this activity we will enhance our intuition about this important measure as well as discover again the surprising relationship between volumes of cylinders (including prisms) and the corresponding cones (pyramids) with same height and base.

Materials: Tape, scissors, a box of rice, unpopped popcorn or Rice Krispies™; copies of the nets (patterns) found on Appendix Pages 22, 23, 24 and 25; and two sheets of $8\frac{1}{2}'' \times 11''$ transparency film

A. Cut out the nets for the open-top prism on Appendix Page 22 and the pyramid on Appendix Page 23. Assemble them carefully, using the tabs to fasten them together.

   (a) Compare the heights and the areas of the bases of the prism and the pyramid.

   (b) Using skills at estimation, compare the volume of the prism to the volume of the pyramid.

   (c) Using the dotted lines, carefully cut a hole in one corner of the bottom of the pyramid. Fill the pyramid with rice (popcorn or Rice Krispies™) and pour it into the prism. Repeat until the prism is full. How many times could you empty a full pyramid into the prism? From this experiment:

   Volume of prism = _____ × Volume of pyramid

B. Cut out the nets for the lateral walls of the cylinder on Appendix Page 24 and the lateral surface of the cone on Appendix Page 25. Assemble them carefully, using the tabs to fasten them together.

   (a) Compare the heights and the areas of the bases of the cylinder and the cone.

   (b) Using your skills of estimation, compare the volume of the cylinder to the volume of the cone.

   (c) Fill the cone with rice (popcorn or Rice Krispies™) and pour it into the cylinder (standing in a shoe box). Repeat until the cylinder is full. How many times can you empty a full cone into the cylinder? From this experiment:

   Volume of cylinder = _____ × Volume of cone

C. Take two sheets of $8\frac{1}{2}$" $\times 11$" transparency paper. Using one sheet, make the lateral surface of a cylinder by taping the short edges together. Using the other sheet, make the lateral surface of a cylinder by taping the long edges together. Stand them side-by-side in a box lid. (Note that the lateral surface area for each cylinder is the same.)

(a) Make a conjecture as to which has the greater volume or if they have the same volume.

(b) To test your conjecture, place the taller of the two cylinders inside the shorter. Fill the taller cylinder with unpopped popcorn. Then, slowly pull the taller cylinder upward. How do the volumes of the two cylinders compare?

(c) Use your knowledge of formulas related to circles and cylinders to determine the volume of each cylinder in cubic inches.

Activity 12-13: Tantalizing Tangrams III - Perimeter

Overview: It is assumed that you will have experienced the joys of making tangrams from a square region (Activity 11-11) and exploring with tangrams (Activity 11-12) prior to this activity. In fact, if you have never made tangrams before, you should work through Activity 11-11 before completing this activity. Tangrams offer a nice connection between experiential geometry and work with irrational numbers. It is further assumed that you know that a 45°- 45°- 90° triangle has a ratio of edges $1:1:\sqrt{2}$ and that you can find the lengths of the edges of these triangles if given the length of one edge.

Materials: A set of tangrams (easily cut with the Ellison Letter Machine™ or from Appendix Page 18)

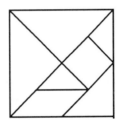

A. Assume your set of tangrams was made from an 8 × 8 square. Find the exact length of each edge of each tangram piece. Then, find the sum of the perimeters of the seven tangram pieces.

Tangram Piece	Perimeter
Small triangle (2)	
Medium triangle	
Parallelogram	
Square	
Large triangle (2)	
SUM of PERIMETERS	

B. If you made your set of tangrams from a 12×12 square, find the exact lengths of the edges of each tangram piece. Find the sum of the perimeters of the seven tangram pieces and record below.

Tangram Piece	Perimeter
Small triangle (2)	
Medium triangle	
Parallelogram	
Square	
Large triangle (2)	
SUM of PERIMETERS	

If your set of tangrams were to be made from an $n \times n$ square, find the exact lengths of the edges of each tangram piece. Find the sum of the perimeters of the seven tangram pieces to generalize this problem.

Tangram Piece	Perimeter
Small triangle (2)	
Medium triangle	
Parallelogram	
Square	
Large triangle (2)	
SUM of PERIMETERS	

TEACHING NOTES: A SUMMARY

Whenever classroom experiences focus on measuring "real" objects which are natural to a student's environment, then students can see the usefulness and necessity of measurement in everyday life. When children serve their own plates at a family gathering, their eyes are often "bigger than their tummies." When they insist on two hamburgers, Biggie fries, and a large shake, they may not be able to hold it all. Anyone putting leftovers in the refrigerator needs to estimate how large a container to take out of the cabinet. We are constantly estimating how far we can drive on a tank of gas, how long it will take to get to school, how many calories are in that wonderful chocolate bar, how long we can stay out before getting sunburned, or how much money to take on a trip. We are constantly estimating in the everyday measurements in which we engage. Therefore, it is imperative that we teachers integrate estimation activities throughout measurement.

- Give balloons to students and ask them to blow up the balloon until it is 12 inches around. (This is an extension of Activity 12-2.)
- Secure syringes (without the needles) which are marked in cubic centimeters (cc's) or milliliters; the ones used for allergy shots are marked to tenths of a cc. Let students fill up syringes to the 1 cc mark and then squirt the water in their palms to *experience* a milliliter.
- Ask students to close their eyes and hold up a finger when they think that a minute has elapsed.
- Investigate how many crisp new dollar bills it would take to weigh a pound or how far one million dollar bills would stretch if laid end-to-end.

You might even involve your whole school in a Great American Popcorn Project and actually count one million pieces of popcorn. (White Bluff Elementary School and Pooler Elementary School in Savannah, GA, did just that.) We integrated science and social studies into this measurement unit. We measured the percentage increase in volume from unpopped kernels to popped; we graphed the number of pieces in a handful from students in kindergarten through fifth grades and then determined the average number of pieces in a handful. We determined the area of the classroom floor in pieces of popped corn; we measured the distance to the principal's office in popcorn pieces. We determined how many students it would take eating popcorn for 30 seconds to eat a million pieces. We wore out three air poppers. Then, we actually got to SEE one million pieces of popcorn at one time!

In addition to the activities in this chapter, be sure to also investigate Activity 7-5 in which the relationship between circumference and diameter of circles is discussed.

As you study this material keep in mind the two-part measurement standard from the NCTM publication *Principles and Standards for School Mathematics:*

MEASUREMENT STANDARD

Instructional programs from prekindergarten through grade 12 should enable all students to—

- *Understand measurable attributes of objects and the units, systems, and processes of measurement*

In prekindergarten through grade 2 all students should–
  - recognize the attributes of length, volume, weight, area, and time;
  - compare and order objects according to these attributes;
  - understand how to measure using nonstandard and standard units;
  - select an appropriate unit and tool for the attribute being measured.

In grades 3–5 all students should–
  - understand such attributes as length, area, weight, volume, and size of angle and select the appropriate type of unit for measuring each attribute;
  - understand the need for measuring with standard units and become familiar with standard units in the customary and metric systems;
  - carry out simple unit conversions, such as from centimeters to meters, within a system of measurement;
  - understand that measurements are approximations and how differences in units affect precision;
  - explore what happens to measurements of a two-dimensional shape such as its perimeter and area when the shape is changed in some way.

In grades 6–8 all students should–
  - understand both metric and customary systems of measurement;
  - understand relationships among units and convert from one unit to another within the same system;
  - understand, select, and use units of appropriate size and type to measure angles, perimeter, area, surface area, and volume.

- *Apply appropriate techniques, tools, and formulas to determine measurements.*

In prekindergarten through grade 2 all students should–
  - measure with multiple copies of units of the same size, such as paper clips laid end to end;
  - use repetition of a single unit to measure something larger than the unit, for instance, measuring the length of a room with a single meterstick;
  - use tools to measure;
  - develop common referents for measures to make comparisons and estimates.

In grades 3–5 all students should–
  - develop strategies for estimating the perimeters, areas, and volumes of irregular shapes;
  - select and apply appropriate standard units and tools to measure length, area, volume, weight, time, temperature, and the size of angles;
  - select and use benchmarks to estimate measurements;
  - develop, understand, and use formulas to find the area of rectangles and related triangles and parallelograms;
  - develop strategies to determine the surface areas and volumes of rectangular solids.

In grades 6–8 all students should–
  - use common benchmarks to select appropriate methods for estimating measurements;
  - select and apply techniques and tools to accurately find length, area, volume, and angle measures to appropriate levels of precision;
  - develop and use formulas to determine the circumference of circles and the area of triangles, parallelograms, trapezoids, and circles and develop strategies to find the area of more-complex shapes;
  - develop strategies to determine the surface area and volume of selected prisms, pyramids, and cylinders;
  - solve problems involving scale factors, using ratio and proportion;
  - solve simple problems involving rates and derived measurements for such attributes as velocity and density.

Review these comments from the narrative of the NCTM *Principles and Standards for School Mathematics:*

Measurement is one of the most widely used applications of mathematics. It bridges two main areas of school mathematics—geometry and number. Measurement activities can simultaneously teach important everyday skills, strengthen students' knowledge of other important topics in mathematics, and develop measurement concepts and processes that will be formalized and expanded in later years.

Teaching that builds on students' intuitive understandings and informal measurement experiences helps them understand the attributes to be measured as well as what it means to measure. A foundation in measurement concepts that enables students to use measurement systems, tools, and techniques should be established through direct experiences with comparing objects, counting units, and making connections between spatial concepts and number.

Students bring to the middle grades many years of diverse experiences with measurement from prior classroom instruction and from using measurement in their everyday lives. In the middle grades, students should build on their formal and informal experiences with measurable attributes like length, area, and volume; with units of measurement; and with systems of measurement.

Measurement concepts and skills can be developed and used throughout the school year rather than treated exclusively as a separate unit of study. Many measurement topics are closely related to what students learn in geometry. In particular, the Measurement and Geometry Standards span several important middle-grades topics, such as similarity, perimeter, area, volume, and classifications of shape that depend on side lengths or angle measures. Measurement is also tied to ideas and skills in number, algebra, and data analysis in such topics as the metric system of measurement, distance-velocity-time relationships, and data collected by direct or indirect measurement. Finally, many measurement concepts and skills can be both learned and applied in students' study of science in the middle grades.

RESOURCES

Periodicals:

"A Triangle Divided: Investigating Equal Areas." *Mathematics Teacher* 93 (October 2000): 608-611.

Battista, Michael, and Douglas C. Clements. "Finding the Number of Cubes in Rectangular Cube Buildings." *Teaching Children Mathematics* 4 (January 1998): 258-264.

Friederwitzer, Fredda J., and Barbara Berman. "The Language of Time." *Teaching Children Mathematics* 6 (December 1999): 254-259.

Hartman, Christina and Patricia. A. Trafton. "Exploring Area with Geoboards." *Teaching Children Mathematics* 4 (October 1997): 72-75-.

"Integrating Measurement Projects: Sand Timers." *Teaching Children Mathematics* 7 (November 2000): 132-135.

"It's All in the Area." *The Mathematics Teacher* 92 (November 1999): 670-672.

Jamski, William D. "Six Hard Pieces." *Arithmetic Teacher* 37 (October 1989): 34-35. *An informal proof of square area with tangrams.*

Kenney, Patricia Ann, and Vicky L. Kouba. "What Do Students Know about Measurement?" In *Results from the Sixth Mathematics Assessment of the National Assessment of Educational Progress,* edited by Patricia Ann Kenney and Edward A. Silver, pp. 141–63. Reston, Va.: National Council of Teachers of Mathematics, 1997.

Kim, Ok-Kyeong, P. Mark Tayler, Ken Simms, and Robert E. Reys. "Do Your Students Measure up Metrically?" *Teaching Children Mathematics* 7 (January 2001): 282-287.

Kubota-Zarivnij, Kathy. "How Do You Measure a Dad?" *Teaching Children Mathematics* 6 (December 1999): 260-264, 251.

LaSaracina, Barbara A., and Sharon K. White. " The Restless Rectangle and the Transforming Trapezoid." *Teaching Children Mathematics* 5 (February 1999): 336-337, 366.

Linquist, Mary. "Implementing the Standards: the Measurement Standards." *Arithmetic Teacher* 34 (May 1987): 34-35.

Malloy, Carol E. "Perimeter and Area through the van Hiele Model." *Mathematics Teaching in the Middle School* 5 (October 1999): 87-90.

"Math-o'-Lanterns." *Teaching Children Mathematics* 6 (October 1999): 72-76.

Moore, Deborah A. "Some Like It Hot: Promoting Measurement and Graphical Thinking by Using Temperature." *Teaching Children Mathematics* 5 (May 1999): 538-543.

"Perimeter or Area? Which Measure Is It?" *Mathematics Teaching in the Middle School* 5 (September 1999): 20-23.

Usnick, Virginia E., Patricia M. Lamphere, and George W. Bright. "A Generalized Area Formula." *Mathematics Teacher* 85 (December 1992): 752-754.

Wilcock, Douglas. "Start the Year Right--Discover Pick's Theorem." *Mathematics Teacher* 85 (September 1992): 424-425.

Woodward, Ernest, and Thomas R. Hamel. "The Use of Dot Paper in Geometry Lessons." *Mathematics Teacher* 86 (October 1993): 558-561.

Young, Sharon L. And Robbin O'Leary. "Creating Number Scales for Measuring Tools." *Teaching Children Mathematics* 8 (March 2002): 400-405.

Books and Other Literature:

Allen, Pamela. *Mr. Archimedes' Bath*. New York: Lothrop, Lee & Shepard Books, 1980.

Clement, Rod. *Counting on Frank*. Milwaukee, WI: Gareth Stevens Publishing, 1991.

Drean, Tamara, and Randall Sourvney. *Measurement Investigations* (1992). Dale Seymour Publications.

Hoff, Syd. *Lengthy*. New York: G. P. Putnam's Sons, 1964.

Lionni, Leo. *Inch by Inch*. New York: Astor-Honor, 1960.

Myller, Rolf. *How Big Is a Foot?* New York: Dell Publishing, 1990.

Other Resources:

Burns, Marilyn. *Mathematics: With Manipulatives* (1988). Cuisenaire Company of America, Inc. *Geoboards Video*.

13 TRANSFORMATIONS AND ADDITIONAL TOPICS IN GEOMETRY

No longer can we afford to sit idly by while our children move through school without receiving the mathematical preparation appropriate for the twenty-first century. The challenges are clear. The choices are before us. It is time to act. - ***Everybody Counts***

Children get their earliest understanding of congruence by tracing a figure and then sliding, turning, or flipping the tracing to see that the tracing has exactly the same size and shape as a second figure. Mathematicians have long had in their tool box the notion of isometries, transformations of the plane that preserve distance (such as translations, rotations, and reflections). Not too many years ago, we realized that this tool, the isometry, would be a useful way to discuss congruence with young people. Translations were renamed slides, rotations were named turns, reflections were named flips, and teachers had a new and powerful approach to use in teaching geometry. In this chapter look forward to activities exploring slides, flips, and turns using paper tracing, a Mira®, ruler and compass, and the coordinate system. Look forward also to an expansion of your knowledge of tessellations. By looking for plane regions that tessellate the plane, we build our intuition.

The research of the van Hieles and others suggests that children learn and experience geometry through a well-defined developmental hierarchy. Children first learn to identify shapes and only later learn to identify the properties that characterize shapes. They first learn the properties of geometric objects, but only later are able to use those properties to form deductive arguments. Certainly the early geometric education of children should attend to the more fundamental of these tasks. However, it is important that students who are developmentally ready be given the opportunity to think deductively about the geometric concepts they have mastered.

The spirit of these early ventures in deductive thinking should be informal rather than formal. Each early exposure to deductive reasoning should follow systematic exploration using concrete representations of the geometric concepts. It should consist of informal conversation about what general properties the exploration revealed and why those general properties are true. Activity 13-8 in which the Side-Angle-Side congruence relation is explored and Activities 13-9 and 13-10 in which the Pythagorean Theorem is visited are examples of explorations that could lead to informal deductive conversation. Geometric constructions are another type of exploration that naturally leads to informal deductive conversation. The many constructions discussed in *Modern Mathematics* are supplemented by Activity 13-11 in which regular 2^n – gons are constructed, Activity 13-12 in which a Mira® is used to bisect line segments and angles, and Activity 13-13 in which some of the unexpected properties of medians, altitudes, and perpendicular bisectors are investigated.

Activity 13-1: A Developmental Approach to Symmetry and Transformations

Overview: To prepare young people to recognize symmetry and to use transformations as a problem-solving tool, it is important that they complete activities that provide the appropriate foundation. The following activity reviews briefly some of the types of exercises that are important steps in this preparation.

A. **Recognizing Differences:** In (a) and (b) find two shapes that are the same in each row. In (c) and (d), identify the shape that is different in each row.

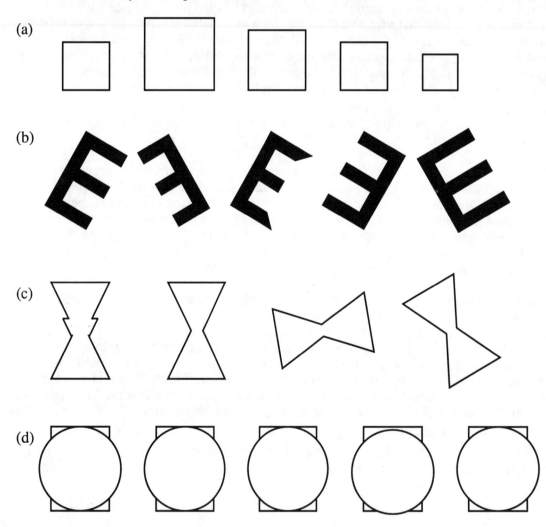

(a)

(b)

(c)

(d)

B. **Identifying Turns:** Each shape without a letter has been obtained by rotating (or turning) a shape with a letter. Find the matching shapes and insert the correct letter.

C. **Identifying Lines of Symmetry:**

(a) Draw a ring around each figure that matches exactly when folded across the line.

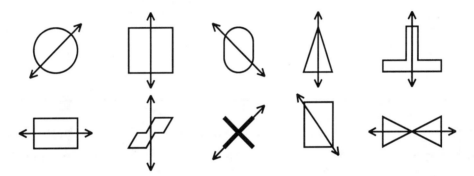

(b) Use a line of symmetry to divide each shape into two shapes that are the same. The first is done for you. If it is possible to do this with more than one line, show all possible solutions.

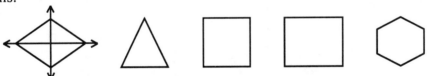

D. **Creating Symmetry:** Complete the following pictures so that the line is a line of symmetry.

(a)

(b)

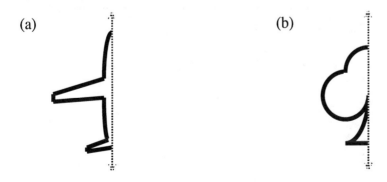

E. **Sketching Reflections:** In each of the following pictures, sketch the picture as it would look reflected in a mirror. The first one is completed for you.

(a)

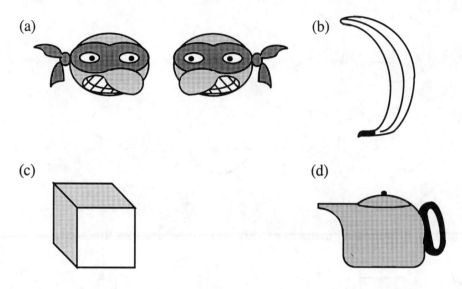

(b)

(c)

(d)

TEACHING   NOTE:

It is important to observe that not every line that divides a region into two congruent parts is a line of symmetry. For instance, in the figure below, line ℓ divides the region into congruent parts but is not a line of symmetry.

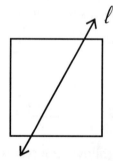

Activity 13-2: Using Tracing Paper to Model Slides, Flips, and Turns

Overview: In this exercise we will use tracing paper to model slides (translations), turns (rotations), and flips (reflections). In the process, we will give careful attention to the properties preserved by these transformations. *Note: In this discussion we will use use prime notation to represent the image of a point under a slide, flip or turn. For instance, P' will be used to represent the image of point P.*

Materials: Tracing paper, ruler, and protractor

A. **Modeling a Slide:** Follow the instructions below to model a slide with the distance and direction indicated by the arrow from *A* to *B*.

    (1) Place a sheet of tracing paper over triangle *PQR* and $\overleftrightarrow{AB}$.

    (2) On the tracing paper trace $\overleftrightarrow{AB}$ and points *P, Q, R,* and *A.*

    (3) Slide the paper until point *A* on the tracing paper coincides with point *B* on the page. Make sure that the traced line continues to coincide with the line on the page.

    (4) With your pencil press an indentation into the page through *P, Q,* and *R* on the tracing paper. Label the indentations on the page as *P', Q',* and *R'* .

    (5) Draw triangle *P'Q'R'*. This is the image of triangle *PQR* under the slide (translation).

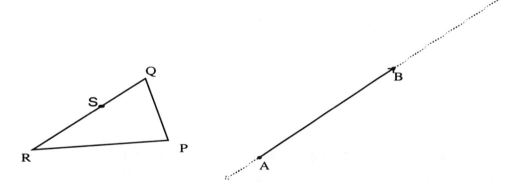

B. **The Properties of a Slide:** Refer to triangle *P'Q'R'* drawn in Part A.

    (a) Measure $\overline{PQ}$.   _____   Measure line segment $\overline{P'Q'}$. _____

    (b) Measure ∠*PQR*.   _____   Measure ∠*P'Q'R'*.   _____

    (c) If we move from point *P* to point *Q* to point *R* in triangle *PQR*, do we move in a clockwise or counterclockwise direction?

    (d) If we move from point *P'* to point *Q'* to point *R* in triangle *P'Q'R'*, do we move in a clockwise or counterclockwise direction?

    (e) *S* is clearly between R and Q on $\overline{RQ}$. Is *S'* between *R'* and *Q'* on $\overline{R'Q'}$?

Note: Our results above suggest that slides preserve distance, angle measure, orientation and betweeness.

C. **Modeling a Turn:** Follow the instructions below to model a turn around point *B* through ∠*ABC*.

(1) Place a sheet of tracing paper over quadrilateral *PQRS* and ∠*ABC*.

(2) On the tracing paper trace ∠*ABC* and points *P, Q, R,* and *S*.

(3) Press firmly on point *B* with your pencil and rotate the tracing paper until point *A* on the tracing paper lies on segment $\overline{BC}$.

(4) With your pencil press an indentation into the page through *P, Q, R,* and *S* on the tracing paper. Label the indentations on the page as *P', Q', R,* and *S'*.

(5) Draw quadrilateral *P'Q'R'S'*. This is the image of quadrilateral *PQRS* under the turn (rotation).

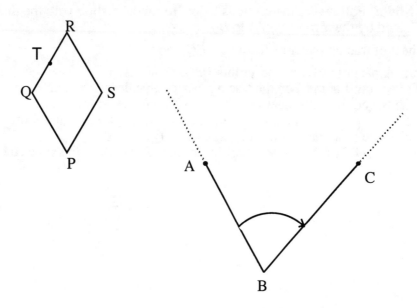

D. **The Properties of Turns:** Refer to quadrilateral *P'Q'R'S'* drawn in part C.

(a) Measure line segment $\overline{PQ}$.   _____   Measure line segment $\overline{P'Q'}$. _____

(b) Measure ∠*PQR*.   _____   Measure ∠*P'Q'R'*.   _____

(c) If we move from point *P* to point *Q* to *R* in quadrilateral *PQRS*, do we move in a clockwise or counterclockwise direction?

(d) If we move from point *P'* to point *Q'* to *R'* in quadrilateral *P'Q'R'S'*, do we move in a clockwise or counterclockwise direction?

(e) Is *T* between *Q'* and *R'* _____   Is *T'* between *Q'* and *R'*? _____

(f) What do these investigations suggest about properties preserved by turns (rotations)?

E. **Modeling a Flip:** Follow the instructions below to model a flip over $\overleftrightarrow{AB}$.(Alternatively, if a Mira® is available, use it to model the flip of triangle *PQR*.)

(1) Place a sheet of tracing paper over triangle *PQR*, $\overleftrightarrow{AB}$.

(2) On the tracing paper trace $\overleftrightarrow{AB}$, and points *P*, *Q*, *R* , *A*, and *B*.

(3) Turn the paper over face down. Make sure that the line on the tracing paper coincides with the line on the page and that points *A* and *B* on the tracing paper coincide with points *A* and *B* on the page.

(4) With your pencil press an indentation into the page through *P*, *Q*, and *R* on the tracing paper. Label the indentations on the page as *P'*, *Q'*, and *R'* .

(5) Draw triangle *P'Q'R'*. This is the image of triangle *PQR* under the flip (reflection).

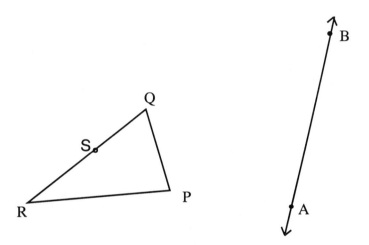

F. **The Properties of Flips:** Refer to triangle *P'Q'R'* drawn in Part E.

(a) Measure line $\overline{PQ}$.   _____   Measure $\overline{P'Q'}$.   _____

(b) Measure ∠*PQR*.   _____   Measure ∠*P'Q'R'*.   _____

(c) Is *S* between R and Q? _____   Is *S'* between R' and Q'?   _____

(d) If we move from point *P* to point *Q* to R in triangle *PQR*, do we move in a clockwise or counterclockwise direction?

(e) If we move from point *P'* to point *Q'* to R in triangle *P'Q'R'* , do we move in a clockwise or counterclockwise direction?

(f) What do (a) and (b) suggest about properties preserved by flips?

Note: The results of (d) and (e) suggest that flips reverse orientation.

Activity 13-3: Looking at Reflections in a Mira®

Overview: In Activity 13-2, we used tracing paper to observe many of the properties of reflections (flips). In particular, we collected evidence that flips preserve distance and angle measure but reverse orientation. In this activity we will find that the Mira® is an excellent tool for studying this family of transformations. The Mira® is a tool built of materials that are transparent (you can see through them) and that reflect. Hence, when you look carefully at one side of a Mira, you see not only a reflection of objects on your side of Mira®, but you also see through the Mira® to the other side.

Materials: Mira®, ruler, protractor, and compass

A. **Experimenting with a** Mira®: Place the Mira® (with beveled edge down and toward you so you can easily draw a line along that beveled edge) between the two images in each pair found below. Adjust Mira® until the reflected image of the figure on the left coincides with the figure seen through the Mira®. Sketch the line. This is called the Mira® line. When the image on the left is flipped across the Mira® line, the image is the figure on the right.

(a)   (b)

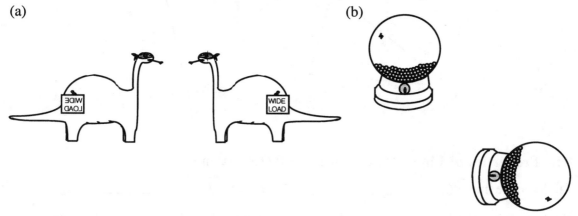

B. **Flips with a** Mira®: Did your work above suggest how to complete flips using the Mira? Place the beveled edge along the line across which you wish to flip the figure. On the other side of the Mira® use a pencil to sketch an image (the flip image) that coincides with the original figure as seen in the Mira®. Use the Mira® to complete the following flips. Label the flip images using prime notation. That is, if point P is in the original figure, label the corresponding point P'.

(a)   (b)

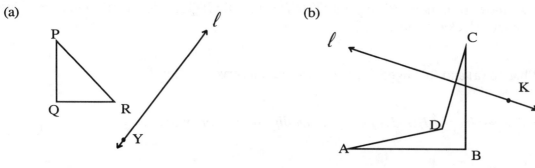

C. **Properties of Flips:** In Part B compare the following measures.

   (a) Length of $\overline{PQ}$ _____   Length of $\overline{P'Q'}$ _____

   (b) Measure of $\angle PQR$ _____   Measure of $\angle P'Q'R'$ _____

   (c) These results from (a) and (b) confirm which properties of flips?

D. **More Properties of Flips:** In Part B (a), sketch $\overline{PP'}$ and in Part B (b), sketch $\overline{AA'}$.

   (a) Let the point of intersection of $\overline{PP'}$ with the flip line be point X. Compute the following measures:

      The length of $\overline{PX}$ _____   The length of $\overline{XP'}$_____

      The measure of $\angle PXY$_____

   (b) Let the intersection of $\overline{AA'}$ and the flip line be point J. Compute the following measures:

      The length of $\overline{AJ}$ _____   The length of $\overline{JA'}$ _____

      The measure of $\angle AJK$ _____

   (c) What do your observations from Part D suggest about the relationship between the flip line and the line segment connecting a point and its image (for instance, what is the relationship between the flip line and $\overline{PP'}$)?

E. **Ruler and Compass Construction of Flips:** In Part D we learned that if P' is the flip image of P over the line ℓ, then ℓ is the perpendicular bisector of $\overline{PP'}$. Use this fact to complete a ruler and compass construction of the flip image of triangle NOP over line ℓ.

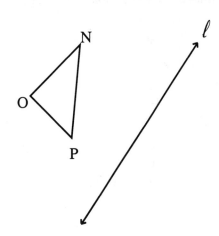

Activity 13-4: Two Flips Make a What!? (Part I)

Overview: In this activity we will investigate a somewhat unexpected relationship between flips and slides.

Materials: Mira®, ruler, and tracing paper

A. Follow these instructions using the figure below.

   (1) Using a Mira® (see Activity 13-3), flip triangle *ABC* across line ℓ_1 and identify the vertices of the flip image using prime notation. (*e.g.,* The image of *A* is denoted *A'* and so on.)

   (2) Using the Mira®, flip triangle *A'B'C'* across line ℓ_2 and identify the flip image using double prime notation. (*e.g.,* The image of *A'* is *A"* and so on.)

   (3) Using tracing paper (see Activity 13-2), find the slide image of triangle *ABC* along an arrow from *A* to *A"*. What is the relationship between the slide image of triangle *ABC* and the image of triangle *ABC* under the double flip (triangle *A"B"C"*)?

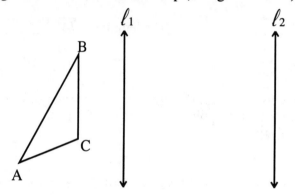

B. Follow the instructions from Part A using the figure below. What do you observe about the relationship between the slide image and the double flip image?

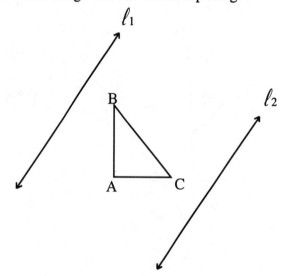

C. (a)   In Part A, measure

the length of the slide (length of $\overline{AA''}$) _____

the distance between ℓ_1 and ℓ_2 _____

 (b) In Part B, measure

the length of the slide (length of $\overline{AA''}$) _____

the distance between ℓ_1 and ℓ_2 _____

 (c) The investigations in Part A and Part B suggest that two successive flips over parallel lines produce the same image as a slide. What is the relationship between the length of the slide and the distance from ℓ_1 to ℓ_2?

Note: Not only do two successive flips over parallel lines produce the same image as a slide, for every slide there are parallel lines so that the slide is equivalent to successive flips over those two lines.

Activity 13-5: Two Flips Make a What!? (Part II)

Overview: In this activity we will investigate a somewhat unexpected relationship between flips and turns.

Materials: Mira®, ruler, protractor, and tracing paper

A. Follow these instructions using the figure below.

(1) Using the Mira® (see Activity 13-3) flip quadrilateral *PQRS* across $\overleftrightarrow{AB}$ and identify the vertices of the flip image using prime notation. (*e.g.,* The image of *P* is denoted *P'* and so on.)

(2) Using the Mira®, flip quadrilateral *P'Q'R'S'* across $\overleftrightarrow{BC}$ and identify the flip image using double prime notation. (*e.g.,* The image of *P'* is *P''* and so on.)

(3) With your ruler, draw ∠*PBP''*. Using tracing paper (see Activity 13-2), draw the turn image of quadrilateral *PQRS* in a turn around *B* through ∠*PBP''*.

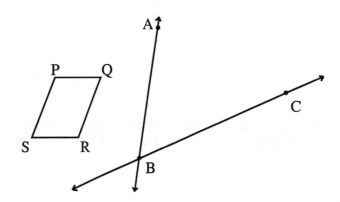

What is the relationship between the turn image of quadrilateral *PQRS* and the image of quadrilateral *PQRS* under the double flip (quadrilateral *P''Q''R''S''*)?

B. Follow the instructions from Part A using the figure below. What do you observe about the relationship between the turn image and the double flip image?

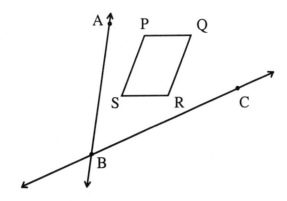

C. (a) In Part A, measure

∠*ABC*   _____

∠*PBP"*   _____

(b) In Part B, measure

∠*ABC*   _____

∠*PBP"*   _____

(c) The investigations in Part A and Part B suggest that two successive flips over intersecting lines produce the same image as a turn. What is the measure of the angle of the turn?

Note: Not only do two successive flips over intersecting lines produce the same image as a turn through an angle with measure twice the measure of the angle between the two lines, for every turn there are intersecting lines so that the turn is equivalent to successive flips over those two lines.

Activity 13-6: Coordinates and Transformations

Overview: Once the coordinate system has been introduced to students in the upper elementary grades, one has available a powerful tool for studying transformations. In the following activities we will discover some patterns that govern transformations in the coordinate system.

Materials: Graph paper (or the square grid paper found on Appendix Page 26), tracing paper, and a Mira®

A. **Reflections Across the y-Axis:** Use a Mira® to reflect the points across the y-axis.

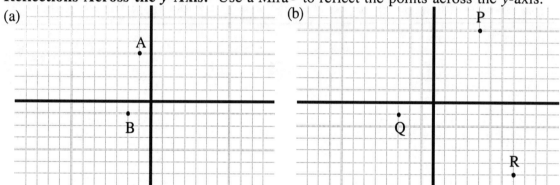

B. For each of the reflections of Part A complete the table listing the coordinates of the points and their images under the reflection across the y-axis.

(a)

	Coordinates of Point	Coordinates of Image
A		
B		

(b)

	Coordinates of Point	Coordinates of Image
P		
Q		
R		

(c) Complete this rule that describes what happens to the coordinates of a point when it is reflected across the y-axis:

When point (x, y) is reflected across the y-axis, the image of (x, y) is _____.

C. **Reflections Across the x-Axis:** Return to the coordinate systems in Part A and use a Mira® to reflect the points in Part A across the x-axis.

D. For each of the reflections of Part C, complete the table giving the coordinates of the points and their images under the reflection about the *x*-axis.

(a)

	Coordinates of Point	Coordinates of Image
A		
B		

(b)

	Coordinates of Point	Coordinates of Image
P		
Q		
R		

(c) Complete this rule that describes what happens to the coordinates of a point when it is reflected across the *x*-axis:

When point (x, y) is reflected across the x-axis, the image of (x, y) is _____.

E. **Translations in a Coordinate System:** Use tracing paper (See Activity 14-2) to translate each figure below in the distance and direction indicated by the arrow.

(a)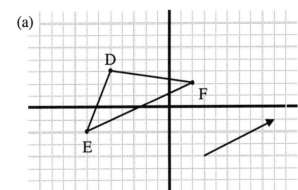

(b)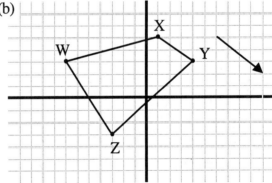

F. (a) The arrow in (a) of Part E points _____ units horizontally and _____ units vertically.

(b) Complete the table giving the coordinates of each point and its image under the translation in (a) of Part E.

	Coordinates of Point	Coordinates of Image
D		
E		
F		

(c) The arrow in (b) of Part E points _____ units horizontally and _____ units vertically.

(d) Complete the table giving the coordinates of each point and its image under the translation in (b) of Part E.

	Coordinates of Point	Coordinates of Image
W		
X		
Y		
Z		

(e) Complete this rule that describes what happens to the coordinates of a point when it is translated:

When point (x, y) is translated along an arrow that points a units horizontally

and b units vertically, the image of the point (x, y) is _____.

Activity 13-7: SAS or SSA?

Overview: The fact that we need only look at a few measurements to determine whether two triangles are congruent is very useful in geometry. In this activity we will explore what information is necessary before we can conclude that two triangles are congruent.

Materials: Protractor, compass, ruler, paper, and scissors

A. **Side-Angle-Side:**

    (a)   On your own sheet of paper, draw two copies of a 35° angle named ∠ABC. The sides of these angles should be at least 10 cm long.

                           b

                           a

        (i)   On your first copy of ∠ABC construct a segment that begins at B, lies along $\overrightarrow{BA}$, and whose length is equal to the length of segment a. Similarly construct a segment whose length is equal to the length of segment b that begins at B and lies along $\overrightarrow{BC}$. Form a triangle by connecting the ends of your segments.

       (ii) On your second copy of ∠ABC construct a segment whose length is equal to the length of segment a that begins at B and lies along $\overrightarrow{BC}$. Construct a segment whose length is equal to the length of segment b that begins at B and lies along $\overrightarrow{BA}$. Form a triangle by connecting the ends of your segments.

      (iii) Cut out the two triangles formed. Place them on top of one another to determine if they are congruent.

      (iv) Can you think of other ways that a 35° angle can be included between segments congruent to segment a and segment b to form a triangle?

    (b)   Repeat (a) using an angle with a different measure and two different lengths of segments.

    (c)   Write a sentence that summarizes what you have learned.

B. **Side-Side-Angle:** On the angle below, locate D on $\overrightarrow{AC}$ where the length of $\overline{AD}$ is equal to the length of segment a from Part A. Set the compass to measure the length of segment b and, with D as center, draw an arc that intersects $\overleftrightarrow{AB}$ in two places. Call these two points of intersection E and F.

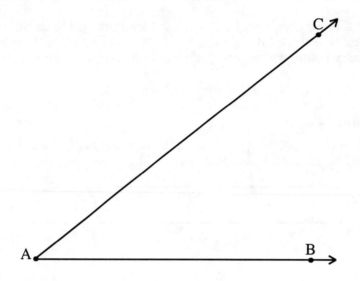

(a) Sketch triangles ADE and ADF. Are these two triangles congruent?

(b) If there is a correspondence between two triangles so that two edges of one triangle are congruent to the corresponding edges of the other triangle and a pair of corresponding angles are congruent, are the triangles necessarily congruent? Explain.

(c) Is there a Side-Side-Angle congruence property?

Activity 13-8: Preparing for Pythagoras

Overview: One of the most famous and important theorems in all of mathematics is the Pythagorean Theorem. Pythagorean triples were known to the Babylonians of Hammurabi's time, more than a thousand years before Pythagoras. However, it is felt by historians that the first general proof of the theorem may well have been given by Pythagoras (Eves, 1990). Today, we use an algebraic statement of the Pythagorean Theorem:

In a right triangle with legs of length a and b and hypotenuse of length c, $a^2 + b^2 = c^2$.

It is important for students to understand the Pythagorean Theorem geometrically. This activity first appeared in the *Mathematics Teacher* (NCTM) in 1979 and again in April 1989. You should have completed Activity 12-6, Activity 12-7, and Activity 12-8 prior to this activity.

Complete this activity in groups of 3 or 4 people.

Materials: *The Theorem of Pythagoras* (See Video Resource List, Appendix Page 32)

A. Similar figures have been constructed on the legs and the hypotenuse of right triangles. Determine the areas of the three similar figures around each right triangle and record the areas in the table provided. Figures A - C are drawn on isometric dot grids, and the basic unit of area is the triangular unit. In Figures D - G, the basic unit of area is the square unit. The areas of the quarter circles in Figure H can be found by using the formula $A = \pi r^2$. As you determine the three areas in each Figure, discuss your strategies with your group.

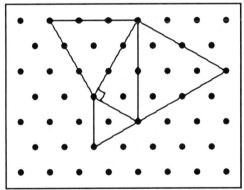
Figure A. Similar equilateral triangles

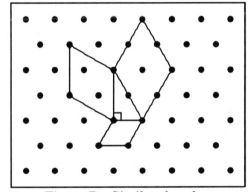
Figure B. Similar rhombuses

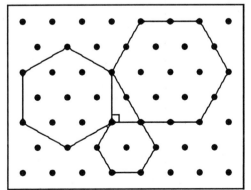
Figure C. Similar hexagons

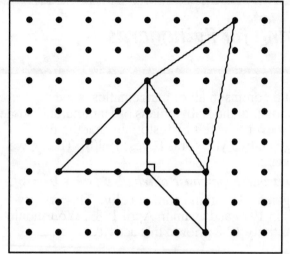

Figure D. Similar right triangles

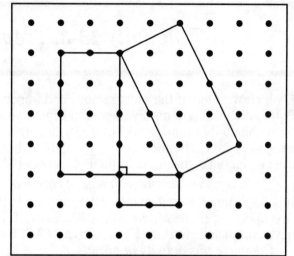

Figure E. Similar rectangles

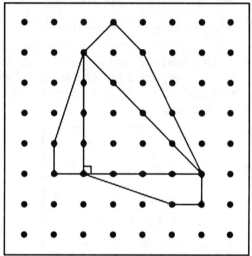

Figure F. Similar trapezoids

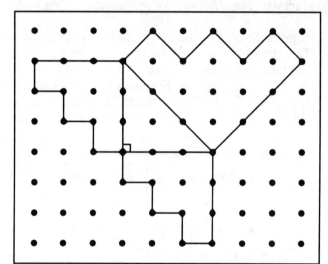

Figure G. Similar octagons

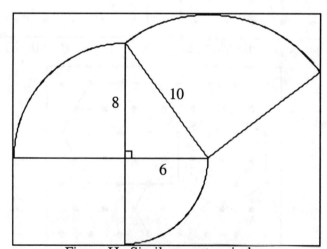

Figure H. Similar quarter circles

Figure	Area of Figure on One Leg	Area of Figure on Other Leg	Area of Figure on Hypotenuse
A			
B			
C			
D			
E			
F			
G			
H			

B. Look at the area values for each figure in the chart above. What relationship do you see among the areas for each figure?

C. Suppose three similar figures are constructed on the edges of a right triangle. The areas of the figures on the two legs are 128 and 450. What is the area of the figure on the hypotenuse?

D. If similar figures are constructed on the edges of a right triangle, then the area of the figure on the hypotenuse is equal to _____.

E. The statement in (d) is the most general statement of the Pythagorean Theorem! Today, we use a specific instance involving squares. For the right triangle below, draw a geometric interpretation for the statement of the Pythagorean Theorem:

In a right triangle with legs of length a and b and hypotenuse of length c, $a^2 + b^2 = c^2$.

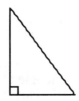

F. View *The Theorem of Pythagoras* from *Project MATHEMATICS!* by Tom Apostol.

Activity 13-9: A Stockbroker Looks at the Pythagorean Theorem

Overview: In 1830, Henry Perigal, a London stockbroker, discovered a way to indicate the truth of the Pythagorean Theorem using paper and scissors. Follow his ideas and determine if they help convince you of the truth of this very important theorem.

Materials: Paper, compass or protractor, ruler, and scissors

A. In the figure below, $\overline{AB}$ is constructed so that it passes through the center of square I and it is perpendicular to the hypotenuse of the right triangle. $\overline{CD}$ is constructed through the center of Square I perpendicular to $\overline{AB}$. Cut out Square II and cut Square I into the four pieces formed by the two segments. Use these five polygonal regions to fill the interior of Square III. Why does this suggest the truth of the Pythagorean Theorem?

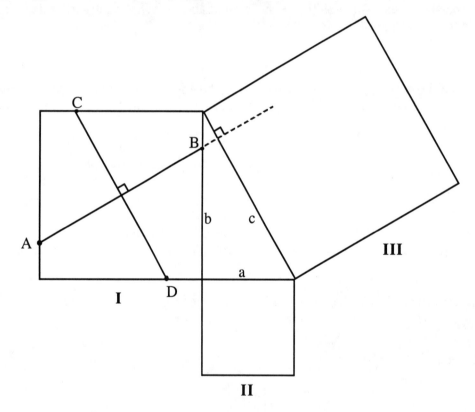

B. On your own sheet of paper draw a right triangle of your choice. On the legs, draw Squares I and II. On the hypotenuse, draw square III. Partition the larger of Squares I and II into four parts using lines as described in (a). Demonstrate the truth of the Pythagorean Theorem for this example.

Activity 13-10: From the Pythagorean Theorem to the Distance Formula

Overview: Among the virtues of the Cartesian coordinate system is the fact that it allows us to compute the distance between any two points in the plane using nothing more sophisticated than the Pythagorean Theorem.

Materials: Graph paper (Note, if graph paper is not available, a Cartesian coordinate system can easily be drawn using the square grid paper on Appendix Page 26)

A. In the coordinate system below, we wish to find the distance between point P with coordinates $(1, 2)$ and Q with coordinates $(5, 7)$. Observe that the dotted segments, together with $\overline{PQ}$, form a right triangle and the length of $\overline{PQ}$ is the hypotenuse of the right triangle. Compute the distance between P and Q using the Pythagorean Theorem.

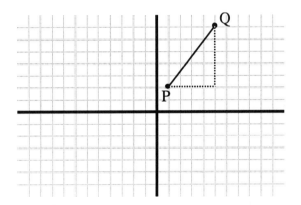

(a) Length of the horizontal leg of the triangle: _____

(b) Length of the vertical leg of the right triangle: _____

(c) Distance between P and Q: _____

B. Place two points A and B of your choice on a coordinate system, sketch the right triangle with $\overline{AB}$ as the hypotenuse, and use the Pythagorean Theorem to compute the distance from A to B.

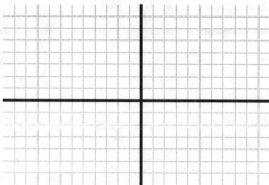

(a) Length of the horizontal leg of the triangle: _____

(b) Length of the vertical leg of the right triangle: _____

(c) Distance between A and B: _____

C. Consider the points A with coordinates (x_1, y_1) and B with coordinates (x_2, y_2) on the coordinate system below.

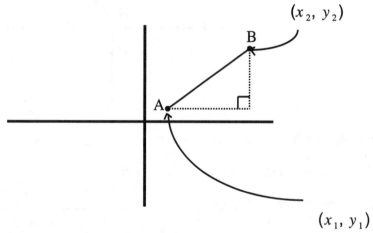

(a) What is the length of the horizontal leg of the right triangle?   _____

(b) What is the length of the vertical leg of the right triangle?   _____

(c) What is the distance from point A to point B?   _____

ACTIVITY 13-11: Constructing Regular 4-, 8-, 16-, ..., 2^n-Gons

Overview: In this activity we will construct a square and then observe that it is easy to construct a regular octagon, 16-gon, 32-gon, and so on.

Materials: Ruler and compass.

A. **Construct a square:** Construct a pair of perpendicular lines that intersect in point X. Connect the four points of intersection of the lines and the circle. The resulting quadrilateral is a square. How do you know?

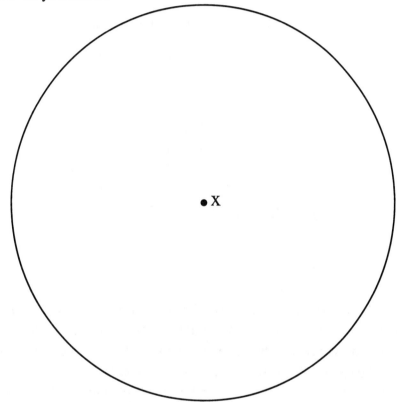

B. **Construct a regular octagon:** Name the four vertices of the square from Part A as A, B, C, and D. Bisect $\angle AXB$. Let the point of intersection of the angle bisector and the circle be E. Set your compass to span $\overline{AE}$. Starting with A, use this compass setting to mark off eight points (including the 4 vertices of the square) around the circle. Connect these points to form a regular octagon.

C. **Construct a regular 16-gon:** Bisect $\angle AXE$ from the octagon formed in Part B. Let the angle bisector intersect the circle at F. Measure arc AF with your compass and use it to mark 16 points around the circle starting with A. Connect these points to create a regular 16-gon.

D. Describe in words how to create a regular 32-gon.

ACTIVITY 13-12: Bisecting with a Mira®

Overview: The Mira® or Image Reflector is a tool built of materials that are transparent (you can see through it) and that reflect. Hence, when you look carefully at one side of a Mira®, you see not only a reflection of objects on your side of Mira®, but you also see through the Mira® to the other side. The Mira® was used in the previous chapter to study reflection transformations. However, we shall see in this activity that it can also be used to bisect line segments and angles.

Materials: A Mira® or Image Reflector and a sharp pencil (For Part D you need a compass)

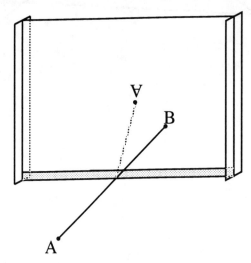

A. Observe that one edge of the Mira® is beveled. The bevel is to facilitate drawing lines; the beveled edge should be down and the bevel should face you. To bisect $\overline{AB}$, place the Mira® across $\overline{AB}$. Adjust the Mira® until the reflected image of A coincides with B seen through the Mira®. (Note: In the picture above, the Mira® needs to be adjusted slightly until the reflected image of A coincides with B.) Carefully draw a line segment along the beveled edge of the Mira®. This line segment is a perpendicular bisector of $\overline{AB}$. Use your Mira® to find the perpendicular bisector of $\overline{CD}$.

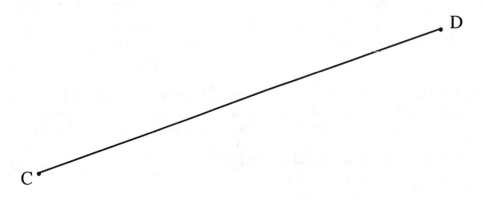

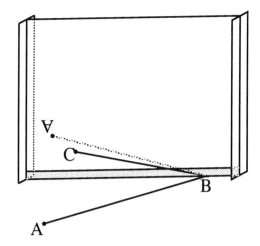

B. The Mira® can also be used to bisect ∠*ABC*. Place the beveled edge of the Mira® so that it lies on the vertex *B*. Adjust the Mira® (keeping the edge on *B*) until the reflected image of $\overline{AB}$ coincides with $\overline{BC}$ seen through the Mira®. Carefully draw in the angle bisector. Now, use the Mira® to bisect ∠*PQR* below.

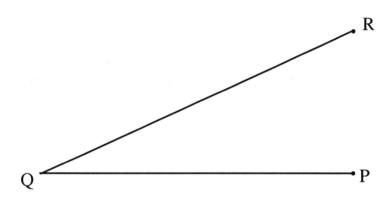

C. The Mira® can also be used to find a perpendicular from point P to $\overline{CD}$. Place the beveled edge on P and across $\overline{CD}$. Adjust the Mira® (keeping the edge on P) until the reflected portion of the line segment coincides with the other portion of the line segment. Use the Mira® to construct a perpendicular segment through P to $\overline{CD}$.

P.

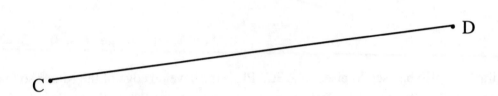

D. Find the intersections of the three bisectors of the interior angles of this triangle. Are you surprised that all three bisectors pass through the same point? This point is called the **incenter** and is the center of the inscribed circle of the triangle (a circle that is tangent to each of the edges). Use a compass to construct this circle.

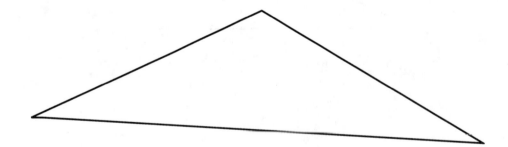

Activity 13-13: Exploring with Medians, Altitudes, and Perpendicular Bisectors

Overview: Connected to each triangle are three important sets of lines or segments: **Median:** A segment connecting a vertex to the midpoint of the opposite edge. **Altitude:** A perpendicular segment from a vertex to the opposite edge (possibly extended). **Perpendicular Bisector:** A line or segment that both bisects the edge and is perpendicular to it. Somewhat unexpected results follow when we experiment with these segments or lines. In this activity we will construct all medians of a triangle and look at the points of intersection. Similarly, we will explore altitudes and perpendicular bisectors. We recommend that you do these constructions with a Mira®, but if this tool is not available, you may perform the constructions with ruler and compass. (If you are using a Mira®, do Activity 13-13 before completing this activity.)

Materials: Mira® (or compass and straight edge if a Mira® is not available) and three different colors of pencils. (If you do Part D of this exercise, you will also need a straight edge and compass, scissors, and cardstock.)

A. Using one of your colored pencils, construct perpendicular bisectors of the edges of the triangles found below. For each triangle, extend these bisectors until they intersect.

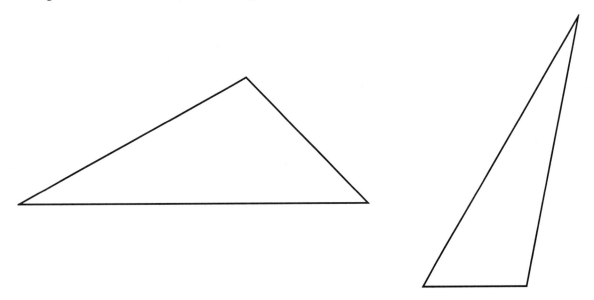

   In view of this exploration, what might you conjecture about the intersection of the perpendicular bisectors of the edges of a triangle?

B. Using the second colored pencil, construct the three medians of each triangle. In view of this exploration, what might you conjecture about the intersection of the medians of a triangle?

C. Using the third colored pencil construct the three altitudes of each triangle. Extend the altitudes of each triangle until they intersect. In view of this exploration, what might you conjecture about the intersection of altitudes of a triangle?

D. In a triangle the *common point* of intersection of the:
- perpendicular bisectors of the edges is called the **circumcenter**.
- medians of the triangle is called the **centroid**.
- altitudes of the triangle is called the **orthocenter**.

(a) The circumcenter is the center of the circumscribed circle that passes through all the vertices of the triangle. Use a compass and sketch this circumscribed circle.

(b) The centroid is the balance point or center of gravity of the triangle. Cut a copy of one of the triangles in Part A from cardstock and observe that if you place your finger under the centroid, the triangle will balance on the end of your finger.

(c) If you constructed the circumcenter, the centroid, and the orthocenter carefully, they should be collinear. Use your ruler and sketch a line that passes through each of these points. This line that contains all three points is called the **Euler line**.

Activity 13-14: To Tessellate or Not ... That Is the Question

Overview: A region bounded by a simple closed curve **tessellates** the plane if one can cover the plane with congruent copies of the region so that there are no gaps and no overlaps. In this activity we will investigate whether some favorite polygonal regions will tessellate the plane.

Materials: Sheets of paper and a set of regular *n*-gons (*n* = 3 to 10) made of poster board or plastic (using Appendix Page 17 or cutting with Ellison Letter Machine™.)

A. Using the regular *n*-gons as patterns, attempt to tessellate the plane. As you experiment, divide the regular polygons into two sets: a set that will tessellate the plane and a set that will not tessellate the plane.

Regular Polygons that Tessellate	Regular Polygons that do not Tessellate

B. We can give a deductive argument to support our inductive observations from Part A if we compute the measure of the interior angles of regular *n*-gons. Remember that the sum of the measures of the interior angles of an *n*-gon is given by

$$(n-2) \times 180$$

Use this fact to compute the measure of each interior angle of the regular *n*-gons and record the results in the table below:

Number of Edges	3	4	5	6	7	8	9	10
Sum of Measures of Interior Angles								
Measure of Each Interior Angle								

Observe that as *n* becomes larger the angle measure becomes _____.

C. (a) Equilateral Triangles will tessellate the plane as illustrated below.

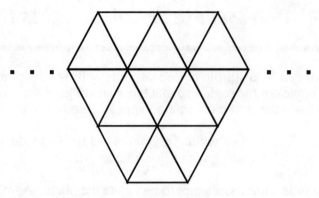

How many triangles intersect at each vertex? _____

What is the sum of the measures of the interior angles that intersect at a vertex? _____

(b) Complete the tessellation of the plane using squares.

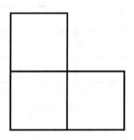

How many squares intersect at each vertex? _____

What is the sum of the measures of the interior angles that intersect at a vertex? _____

(c) Sketch a tessellation of the plane using regular hexagons.

How many hexagons intersect at each vertex? _____

What is the sum of the measures of the interior angles that intersect at a vertex? _____

(d) In any tessellation of the plane by regular polygons, what will be the sum of the measures of the interior angles that intersect at a vertex in the tessellation? _____

D. (a) If *three* regular pentagons intersected at a vertex of a tessellation, the sum of the measures of the interior angles would be _____.

If *four* regular pentagons intersected in a vertex of a tessellation, the sum of the measures of the interior angles about the vertex would be _____.

If *five* regular pentagons intersected in a vertex of a tessellation, the sum of the measures of the interior angles would be _____.

Can the plane be tessellated by regular pentagons? Explain!

(b) Explain carefully why regular septagons cannot tessellate the plane.

(c) Explain carefully why regular octagons cannot tessellate the plane.

(d) Explain carefully why a regular n-gon will not tessellate the plane if $n > 6$.

Activity 13-15: Tessellation Treats

Overview: In Activity 13-14 you discovered that the only regular polygons that tessellate the plane are the equilateral triangle, the square, and the hexagon. In this activity we will learn how to modify polygons that tessellate the plane to create an infinite number of other figures that tessellate the plane. The only limit is your imagination.

Materials: Sheet of stiff cardboard, tape, scissors, and an 11" by 14" sheet of paper. (Optional: Colored pencils or markers)

A. **Slide a Nibble:** Cut a square or regular hexagon from the stiff cardboard and identify a pair of edges that are opposite. As in the picture below, outline a "nibble" from one edge, slide it to the opposite edge, and tape it firmly in place. If you have used a translation (a slide) through the arrow connecting midpoints of opposite edges to move the "nibble" (so that it is attached at corresponding points on the opposite edge), you will have created a pattern with which you can tessellate the plane. Merely trace the pattern, then translate and trace again.

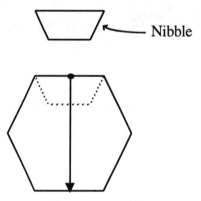

Make a pattern of your own design, trace and tessellate. Decorate as your creative juices lead you!

B. **Double or Triple Nibbles:** Note that in the case of a square you can in fact "nibble and slide to an opposite edge" on each of two adjacent edges and on a hexagon, you can "nibble and slide to an opposite edge" on three adjacent edges. Create a tessellation using a pattern created with two nibbles or three nibbles.

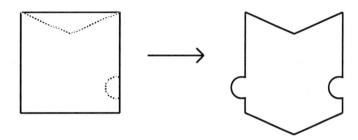

C. **Nibble and Turn:** Starting with a triangle, a square or a hexagon cut from the stiff cardboard, cut a "nibble" from one edge, rotate it around a vertex, and reattach it to the adjacent edge that shares that vertex. If you have applied only a rotation (a turn) about the shared vertex in the process of re-attaching the "nibble," you will have created a pattern that will tessellate the plane. Create a pattern of your own design. By rotating the pattern around the shared vertex and the vertices not involved in the nibble and tracing each new location, tessellate the plane. (Note: In a square or hexagon, you can nibble and turn on each pair of adjacent edges to create a pattern.)

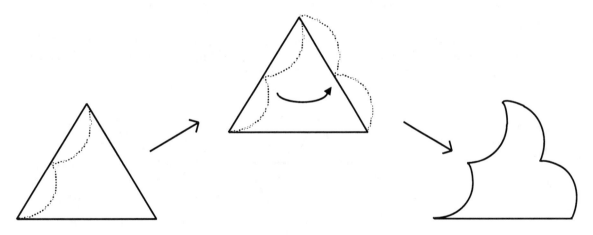

Activity 13-16: Fragments of Fractals

Overview: Over the past two decades researchers have increasingly turned to a new branch of mathematics called **Chaos Theory** when trying to model very complex physical and social phenomena. In 1982 Benoit Mandelbrot introduced the reading public to one of the powerful tools of Chaos Theory when he published the book, *The Fractal Geometry of Nature*. Among the objects discussed in fractal geometry is a set of objects that are described by starting with some relatively simple shape, and then applying a repetitive process called **iteration.** In this activity we will examine one such fractal curve and make some interesting observations about its length and the area it bounds.

A. **The Koch Curve:** The Koch curve is created by starting with an equilateral triangle (See Stage 1 below) and then building a new curve in the following way.

   (i)   Divide each edge into thirds.
   (ii)   Over the middle segment of each edge, erect a new equilateral triangle.
   (iii)   Discard the base of the new equilateral triangle.

   In other words, on each edge replace

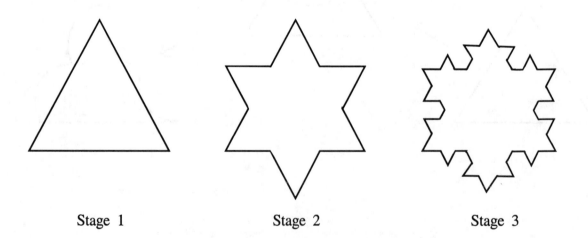

Find the first three stages of the Koch curve below. The Koch curve is the object generated when this iterative procedure is repeated an infinite number of times. Use a pencil and sketch Stage 4 of the Koch curve on the copy of Stage 3 found below.

Stage 1   Stage 2   Stage 3

B. **Length of the Koch Curve**: Let us seek some information about the length of the Koch curve by completing the table on the following page. In this table we will try to find the perimeter of each of the stages of the Koch curve. Assume that the length of each of the edges of the equilateral triangle in Stage 1 is 1.

Stage	Segments per Edge of Original Triangle	Length of Segments	Length of Curve
1	$1 = 4^0$	1	$3(1)(1)$
2	$4 = 4^1$	$\frac{1}{3}$	$(3)(4)(\frac{1}{3})$
3	$16 = 4^2$	$\frac{1}{9}$	$(3)(4^2)(\frac{1}{9})$
4	$64 =$		
n			

Thus, a general formula for the length of the curve in Stage n is _____. If you were to evaluate this formula for larger and larger values of n on a calculator, you would find that the perimeter gets larger and larger without bound. We say that the length of the Koch curve is infinite.

C. **Area Bounded by the Koch Curve:** Can you show that the area of an equilateral triangle of edge length s is given by $\frac{\sqrt{3}}{4} \cdot s^2$?

(Hint: See triangle to the right.)

Now use this fact to complete the following table. In this table we observe that with each new stage of the Koch curve, we add the area of new equilateral triangles to the area bounded by the curve. For instance, in going from Stage 1 to Stage 2 we add three new equilateral triangles, each of which has edges that measure _____.

Stage	Number of New Triangles	Length of Edge of New Triangles	Area of Each New Triangle
1	1	1	$\frac{\sqrt{3}}{4} \cdot 1^2$
2	$3 = 3^1$	$\frac{1}{3}$	$\frac{\sqrt{3}}{4} \cdot (\frac{1}{3})^2$
3	$9 = 3^2$	$(\frac{1}{3})^2$	
4	$27 =$		
n			

From this table it can be seen that the area of Stage n of the Koch curve is given by adding the products of the second column and the fourth column of this table. It can be shown that as n increases, this expression gets closer and closer to a fixed finite number. The Koch curve has an infinite length but bounds a finite area! Wow!

TEACHING NOTES: A SUMMARY

It is crucial that teachers of early elementary students understand that they provide the foundation for mathematical learning that occurs not only in upper elementary school but also in middle school and high school. If a teacher of early elementary students does not focus on this fact, he or she may omit crucial learning objectives that seem trivial or unimportant at the time, but are important pieces in the students' development.

Before students can discuss and use the concepts of symmetry and the problem solving power of transformations of the plane, they must first be able to identify when geometric objects are the same and when they are different and be able to detect what those similarities and differences are. It is important that as students are developmentally ready, they be engaged in activities that develop these powers of discrimination. Activity 13-1 provides an overview of some of the types of paper-and-pencil activities that develop these abilities. If your textbook series does not provide the resources that are needed to develop these geometric critical thinking skills, you might reference *Critical Thinking Activities in Patterns, Imagery, Logic* by Dale Seymour Publications.

It is important, however, that as students begin to work with slides, flips, and turns that their experience be more than paper-and-pencil experience. Pattern blocks are useful manipulatives in this regard. Provide students with a ray and a pattern, and ask them to match the pattern blocks to the pattern, slide the pattern blocks along the ray, and trace the image. Provide a flip line and pattern, ask students to fit pattern blocks in the pattern, flip them across the flip line, and trace the image.

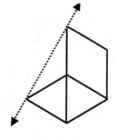

Slide and Trace   Flip and Trace

Have more mature students investigate flips, slides, and turns using tracing paper as in Activity 13-2; using a Mira® as in Activities 13-3, 13-4, 13-5, and 13-6; using ruler and compass as in Activity 13-5; and using coordinate paper as in 13-6.

Tessellations provide an important opportunity to convince the doubters among your students that mathematics can be fun. However, it is important for students to learn and problem solve as they experience the fun. Activity 13-14 invites students to discover which regular polygons tessellate the plane. As students create the "Escher-like" tessellations in Activity 13-15, be sure that they reference the concepts of slide, turn, and flip in their conversation about their work.

RESOURCES

Periodicals:

"Algebra in the Service of Geometry: Can Euler's Line Be Parallel to a Side of a Triangle?" *Mathematics Teacher* 93 (May 2000): 428-431.

Bidwell, James K. "Using Reflections to Find Symmetric and Asymmetric Patterns." *Arithmetic Teacher* 34 (March 1987): 10-15.

Clauss, Judith E. "Pentagonal Tessellations." *Arithmetic Teacher* 38 (January 1991): 52-56.

Enderson, Mary C., Azita Manouchehri, and Lyle A. Pugnucco, "Exploring Geometry with Technology."*Mathematics Teaching in the Middle School* 3 (Mar. - Apr. 1998): 436-42.

Giganti, Paul Jr., and Mary Jo Cittadino. "The Art of Tessellation." *Arithmetic Teacher* 37 (March 1990): 6-16.

Kaiser, Barbara. "Exploring with Tessellating Polygons." *Arithmetic Teacher* 36 (December 1988): 19-24.

Margerm, Pat. "An Old Tale with a New Turn--and Flip and Slide." *Teaching Children Mathematics* 6 (October 1999): 86-90.

"Measuring Up with *The Principal's New Clothes*." *Teaching Children Mathematics* 5 (April 1999): 476-479.

Seidel, Judith, "Symmetry in Season."*Teaching Children Mathematics* 4 (January 1998): 244-49.

Sellke, Donald H. "Geometric Flips via the Arts." *Teaching Children Mathematics* 5 (February 1999): 379-383.

Zaslavsky, Claudia. "Symmetry in American Folk Art." *Arithmetic Teacher* 38 (September 1990): 6-12.

Books and Other Literature:

Bezuszka, Stanley, Margaret Kenney, and Linda Silvey. *Tessellations: The Geometry of Patterns* (1977). Creative Publications.

Britton, Jill, and Walter Britton. *Teaching Tessellating Art* (1992). Dale Seymour Publications.

Ranucci, E.R., and J.L.Teeters. *Creating Escher-Type Drawings* (1977). Creative Publications.

Seymour, Dale. *Critical Thinking Activities in Patterns, Imagery, Logic*. Dale Seymour Publications.

Seymour, Dale, and Jill Britton. *Introduction to Tessellations* (1989). Dale Seymour Publications.

Seymour, Dale. *Tessellation Teaching Masters* (1989). Dale Seymour Publications.

Other Resources:

Apostol, Tom. *Similarity* (1990). *Project MATHEMATICS! Video.*

Apostol, Tom. *The Theorem of Pythagoras* (1988). *Project MATHEMATICS! Video.*

14 ALGEBRA, GEOMETRY, AND GRAPH THEORY

Renowned motto over the door of Plato´s academy:
"Let no one unversed in geometry enter here." - Plato
[from Howard Eves: "Thus, because of its logical element and the pure attitude of mind that he felt its study creates, mathematics seemed of utmost importance to Plato, and for this reason it occupied a valued place in the curriculum of the Academy."]

Those who use mathematics have been presented with no more powerful tool than the coordinate system developed by Rene Descartes in the mid-seventeenth century. The enormous virtue of Descartes' invention is that it allows us to geometrically represent algebraic relationships between pairs of changing quantities, that is, it provides a bridge between algebra and geometry. In recent years the power of the coordinate system has been enhanced enormously by the invention of the graphing calculator.

In the activities of this chapter we will examine some of the most fundamental properties of algebraic relationships represented in the coordinate system: translations and reflections. We will also build an algebraic model of a problem and then use its geometric representation in the coordinate system to solve the problem.

Activity 14-1: Slope on a Geoboard

Overview: The geoboard provides a good physical model of the first quadrant of the Cartesian coordinate system if we regard the lower left corner as the origin, the horizontal edge as the *x*-axis and the vertical edge as the *y*-axis. We can model nonvertical segments of lines on the coordinate system and compute their slopes using the fact that Slope $= \dfrac{\text{Change in } y \text{ coordinates}}{\text{Change in } x \text{ coordinates}}$.

Materials: A geoboard and geobands (*i.e.*, rubber bands)

A. (a) On your geoboard represent the segment seen in the diagram below. Then compute its slope.

   Change in *y* coordinates: _____

   Change in *x* coordinates: _____   Slope: _____

   (b) Represent two additional nonvertical segments on the geoboard and compute their slopes.

   Change in *y* coordinates: _____   Change in *y* coordinates: _____

   Change in *x* coordinates: _____   Change in *x* coordinates: _____

   Slope: _____   Slope: _____

B. (a) On your geoboard represent the segments of parallel lines seen below and compute the slopes of the lines.

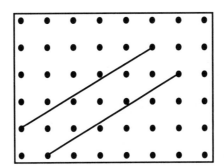

Slope of line 1: _____   Slope of line 2: _____

Represent two additional pairs of nonvertical parallel line segments on the geoboard. Compute the slope of each of the lines represented.

(b) Slope of line 1: _____   Slope of line 2: _____

(c) Slope of line 1: _____   Slope of line 2: _____

(d) From these observations, what conjecture can you make about the slopes of parallel lines?

C. (a) On your geoboard represent the segments of perpendicular lines seen below and compute the slopes of the lines.

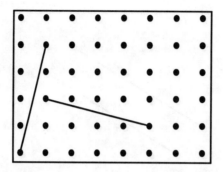

Without using either vertical and horizontal segments, represent two additional pairs of perpendicular line segments on the geoboard. Compute the slope of each of the lines represented.

(b) Slope of line 1: _____   Slope of line 2: _____

(c) Slope of line 1: _____   Slope of line 2: _____

(d) From these observations, what conjecture can you make about the slopes of perpendicular lines?

TEACHING   NOTE:

This activity can be completed using the pictures of 5 x 5 geoboards found on Appendix Page 11 or square dot paper such as that found on Appendix Page 21. The geoboard has the advantage of giving students a tactile experience as part of their problem-solving process; the paper allows students to record their work for future reference. One solution is to do the activity initially on the geoboard and then visit the same types of problems later as a pencil-and-paper exercise or using both media.

Activity 14-2: Translations and the Graphing Calculator

Overview: In Chapter 13 we studied translations of geometric figures in the plane. In this activity we will explore the effects that algebraic modifications of the rule for a function have on the graph of the function. Be sure to use the concept of *translation* when describing those effects.

Materials: A graphing calculator

A. **Adding a Constant:** Graph the following sets of functions on your graphing calculator. Some of these functions may be unfamiliar to you, but their graphs will still be instructive.

   (a) $f(x) = x$ $f(x) = x + 2$ $f(x) = x - 3$

   (b) $f(x) = x^2$ $f(x) = x^2 + 3$ $f(x) = x^2 - 2$

   (c) $f(x) = \dfrac{1}{x}$ $f(x) = \dfrac{1}{x} + 4$ $f(x) = \dfrac{1}{x} - 1$

   (d) Describe in words the relationship between the graph of $f(x)$ and the graph of $f(x) + a$ when $a > 0$. When $a > 0$.

B. **Adding a Constant to the Variable of a Function:** Graph the following sets of functions on your graphing calculator.

   (a) $f(x) = x$ $f(x) = (x - 2)$ $f(x) = (x + 5)$

   (b) $f(x) = \dfrac{1}{x}$ $f(x) = \dfrac{1}{x - 3}$ $f(x) = \dfrac{1}{x + 3}$

   (c) $f(x) = \sin(x)$ $f(x) = \sin(x - 1)$ $f(x) = \sin(x + 3)$

   (d) Describe in words the relationship between the graph of $f(x)$ and the graph of $f(x + a)$ when $a > 0$. When $a < 0$.

C. **Multiplying a Function by a Constant:** Graph the following sets of functions on your graphing calculator.

   (a) $f(x) = x^2$ $f(x) = 3x^2$ $f(x) = \frac{1}{2}x^2$

   (b) $f(x) = \sin(x)$ $f(x) = 4\sin(x)$ $f(x) = \frac{1}{3}\sin(x)$

   (c) Describe in words the relationships between the graph of $f(x)$ and $a \cdot f(x)$ when $a > 1$. When $0 < a < 1$. (Note: The word translation will not be useful here!)

Activity 14-3: Reflections and the Graphing Calculator

Overview: In Chapter 13 we learned about reflections of geometric figures in the plane. In this activity we will explore the effects that algebraic modifications of the rule for a function have on the graph of the function. Be sure to use the word *reflection* when describing these effects.

Materials: A graphing calculator

A. **Multiplying the Function by -1:**   Graph the following pairs of functions on your graphing calculator. Some of these functions may be unfamiliar to you, but their graphs will still be instructive.

(a) $f(x) = 2x + 3$ $f(x) = -(2x + 3)$

(b) $f(x) = x^2$ $f(x) = -x^2$

(c) $f(x) = \ln(x)$ $f(x) = -\ln(x)$

(d) Describe in words the relationship between the graph of $f(x)$ and the graph of $-f(x)$.

B. **Multiplying the Variable (Argument) of the Function by -1:**   Graph the following pairs of functions on your graphing calculator.

(a) $(x) = x$ $f(x) = -x$

(b) $f(x) = x^3$ $f(x) = (-x)^3$

(c) $f(x) = \ln(x)$ $f(x) = \ln(-x)$

(d) Describe in words the relationship between the graph of $f(x)$ and the graph of $f(-x)$.

Activity 14-4: The "Biggest" Rectangle

Overview: A quadratic function is a function of the form $f(x) = ax^2 + bx + c$. $F(x) = x^2$ and $f(x) = 2x^2 + 3x + 1$ are examples of quadratic functions. In this activity we will use quadratic functions to create a model for this problem:

   • *Of all rectangles with a perimeter of 16, which encloses the largest area?*

Then, we will use our model to solve the problem.

Materials: Several 16-inch shoe strings, scissors, ruler, and either a graphing calculator or a piece of graph paper

A. Forming Rectangles:

(a) Cut a pair of 1-inch pieces from one of your shoe strings. Divide the remainder of the shoe string in half and use the four pieces to form a rectangle. We will call the 1-inch edges the *x*-edges of the rectangle.

What is the length of the second edge of the rectangle?_____

What is the area of the rectangle?   _____

(b) From another 16-inch shoe string, form the *x*-edges by cutting a pair of 2-inch pieces. Divide the remainder of the shoe string in half and use the four pieces to form a rectangle.

What is the length of the second edge of the rectangle?_____

What is the area of the rectangle?   _____

(c) Complete the following table in which you record the resulting area when you cut a variety of different lengths to form the x-edges.

Length of x-Edge	Length of Other Edge	Area
1		7
2		12
3		
4		
5		
6		
7		

B. Finding the Quadratic Model:

(a) Look carefully at the table above. What pattern do you see? Write an expression for area when we use a variable x for the x-side.

x	?	?

(b) The expression that you placed in the third rectangle is the rule for the function that describes the area. That is, the area of a rectangle formed from a 16-inch strip with one side of the rectangle equal to x is given by

$$A(x) = \underline{\hspace{2cm}}$$

C. The "Biggest" Rectangle:
Sketch the function $A(x)$ either by entering it into a graphing calculator, or plotting the seven points from the table in Part A [that is, plot (1, 7), (2, 12) ...].

(a) From the graph, what appears to be the largest value that $A(x)$ can obtain? _____

(b) What are the dimensions of the rectangle that has this area?

Note: *Of all possible rectangles with a given perimeter, the square with that perimeter encloses the maximum area.* Do our investigations give evidence for this statement?

TEACHING NOTES: A SUMMARY

The coordinate system itself is introduced and used extensively in the later years of elementary school. However, the study of algebraic relationships using the coordinate system is often postponed until the middle school years and then primary attention is given to linear relationships. In years to come we may see more complex relationships studied earlier using the graphing calculator as a tool.

RESOURCES

Periodicals:

Berman, Barbara, and Fredda Friederwitzer. "Algebra Can Be Elementary ...When It's Concrete." *Arithmetic Teacher* 36 (April 1989): 21-24.

Carlson, Marilyn P. "A Cross-Sectional Investigation of the Development of the Function Concept." In *Research in Collegiate Mathematics Education 3,* edited by Alan H. Schoenfeld, Jim Kaput, and Ed Dubinsky, pp. 114–62. Washington, D.C.: Conference Board of the Mathematical Sciences, 1998.

Demana, Franklin, and Joan Leitzel. "Establishing Fundamental Concepts through Numerical Problem Solving." In *The Ideas of Algebra, K–12,* 1988 Yearbook of the National Council of Teachers of Mathematics, edited by Arthur F. Coxford, pp. 61–68. Reston, Va.: National Council of Teachers of Mathematics, 1988.

Dugdale, Sharon. "Functions and Graphs: Perspectives on Student Thinking." In *Integrating Research on the Graphical Representation of Functions,* edited by Thomas Romberg, Elizabeth Fennema, and Thomas Carpenter, pp. 101–29. Hillsdale, N.J.: Lawrence Erlbaum Associates, 1993.

English, Lyn D., and Elizabeth A. Warren. "Introducing the Variable through Pattern Exploration." *Mathematics Teacher* 91 (February 1998): 166–70.

Kieran, Carolyn. "Helping to Make the Transition to Algebra." *Arithmetic Teacher* 38 (March 1991): 49-51.

Owens, John, E. "Families of Parabolas." *Mathematics Teacher* 85 (September 1992): 477-479.

Schultz, James E. "Teaching Informal Algebra." *Arithmetic Teacher* 38 (May 1991): 34-37.

Smith, Erick. "Patterns, Functions, and Algebra." In *A Research Companion to NCTM's Standards,* edited by Jeremy Kilpatrick, W. Gary Martin, and Deborah Schifter. Reston, Va.: National Council of Teachers of Mathematics, 2002.

Wagner, Sigrid, and Sheila Parker. "Advancing Algebra." In *Research Ideas for the Classroom, High School Mathematics,* edited by Patricia S. Wilson, pp. 119–39. New York: Macmillan Publishing Co., 1993.

Yerushalmy, Michal, and Judah L. Schwartz. "Seizing the Opportunity to Make Algebra Mathematically and Pedagogically Interesting." In *Integrating Research on the Graphical Representation of Functions,* edited by Thomas A. Romberg, Elizabeth Fennema, and Thomas P. Carpenter, pp. 41–68. Hillsdale, N.J.: Lawrence Erlbaum Associates, 1993.

Books and Other Literature:

Bennett, Albert, Eugene Maier, and Ted Nelson. *Math and the Mind's Eye.* Portland, Oreg.: The Math Learning Center, 1988–99.

Educational Development Center, Inc. *Using Algebraic Thinking: Patterns, Numbers, and Shapes.* Mathscape: Seeing and Thinking Mathematically series. Mountain View, Calif.: Creative Publications, 1998.

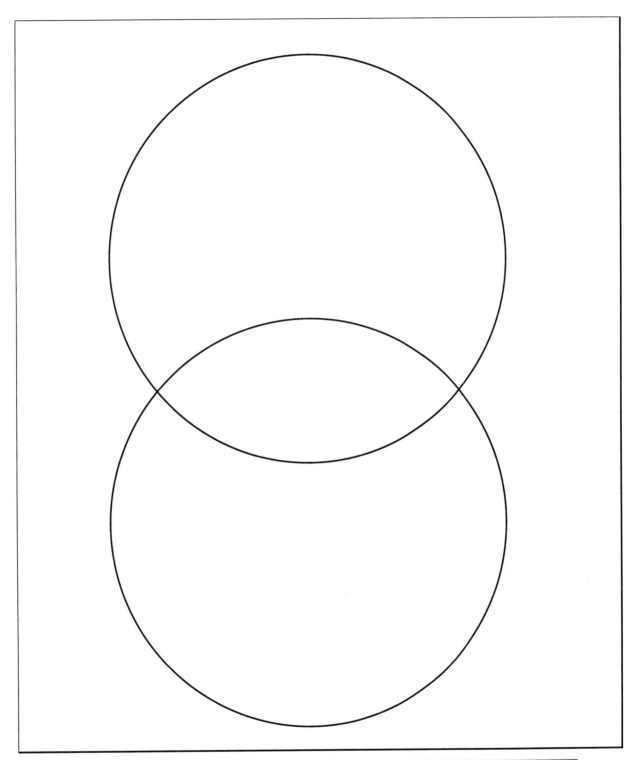

Labels:

	Red	Square	Small
Yellow	Blue	Rhombus	Large
	Green	Triangle	Circle

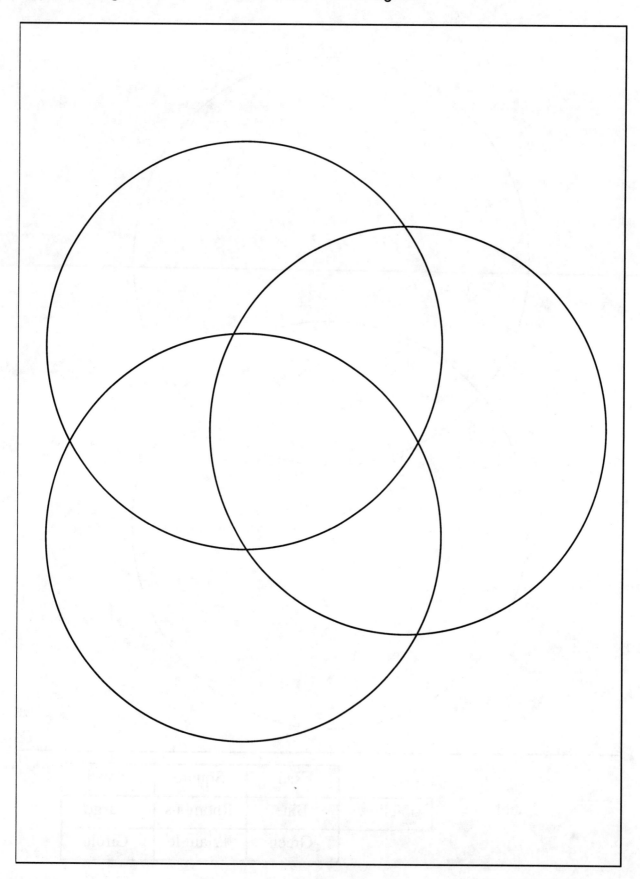

Hundreds	Tens	Ones

	G
	R
	B
	Y

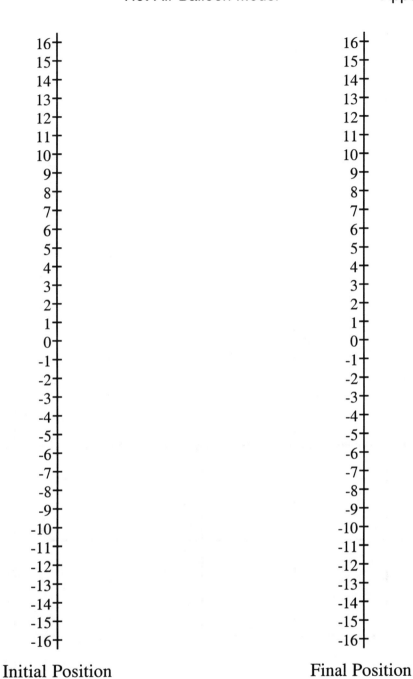

Initial Position   Final Position

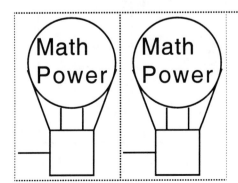

Cut out the balloons and use paper
clips to model arithmetic
operations.

Inch Square Grid

The Circle Fraction Model is available commercially and can be cut out easily using the Ellison Letter Machine.™ It consists of 9 different colored circles cut into various unit fraction sizes. Blackline masters for copying and cutting your own Circle Fraction Model are found below.

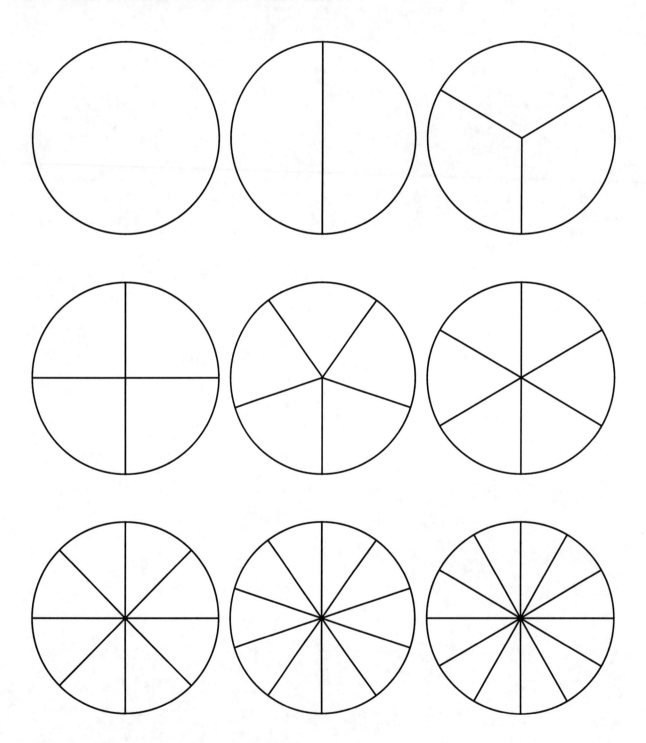

The Square Fraction Model is available commercially and can be cut out easily using the Ellison Letter Machine.™ It consists of 9 different colored squares cut into various unit fraction sizes. Blackline masters for copying and cutting your own Square Fraction Model are found below.

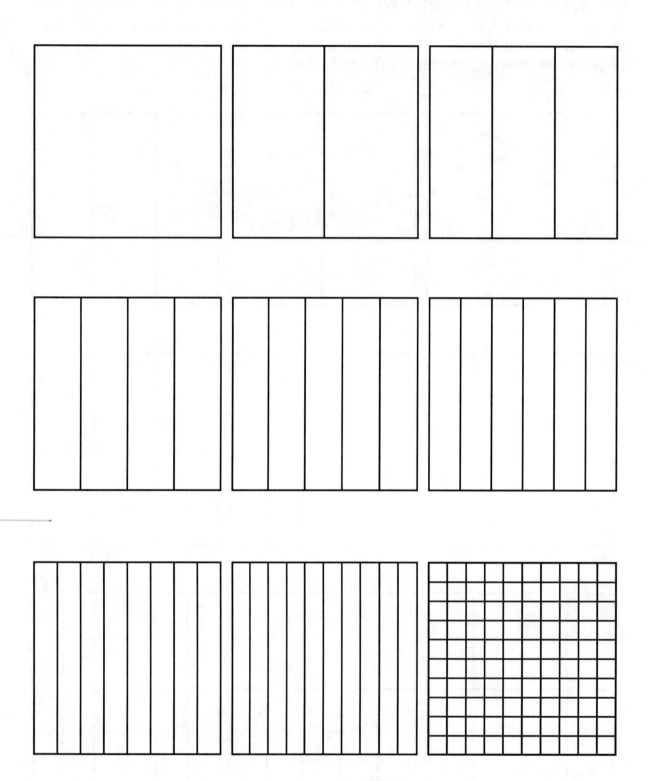

100-Grids

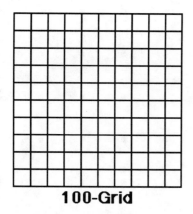

100-Grid

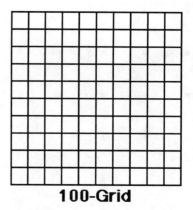

100-Grid

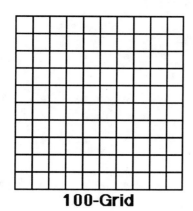

100-Grid

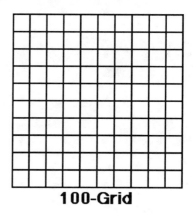

100-Grid

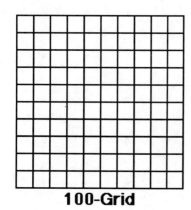

100-Grid

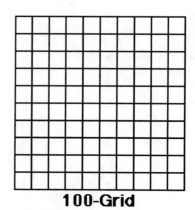

100-Grid

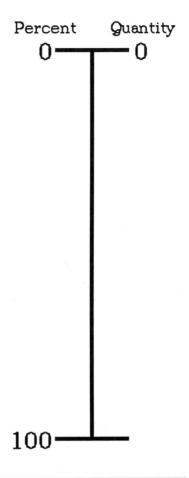

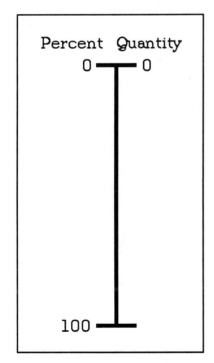

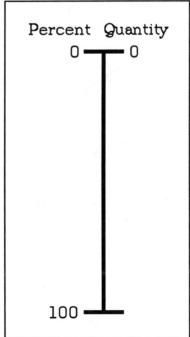

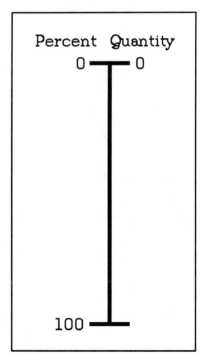

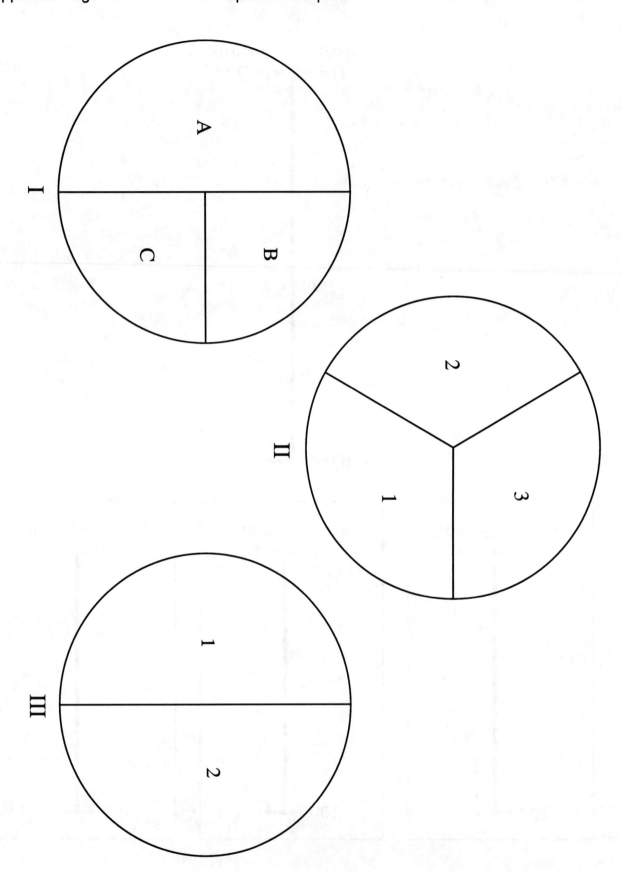

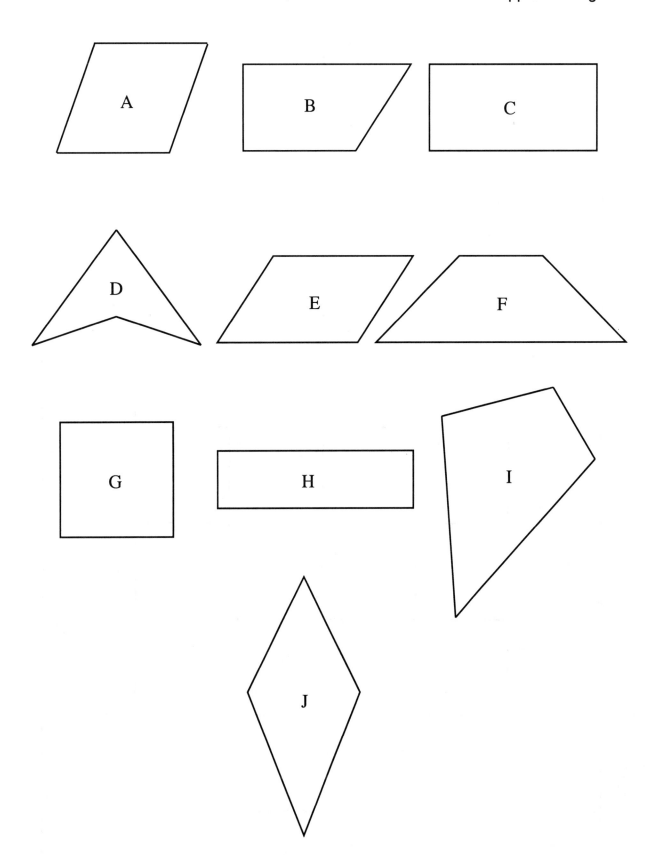

Pentominoes

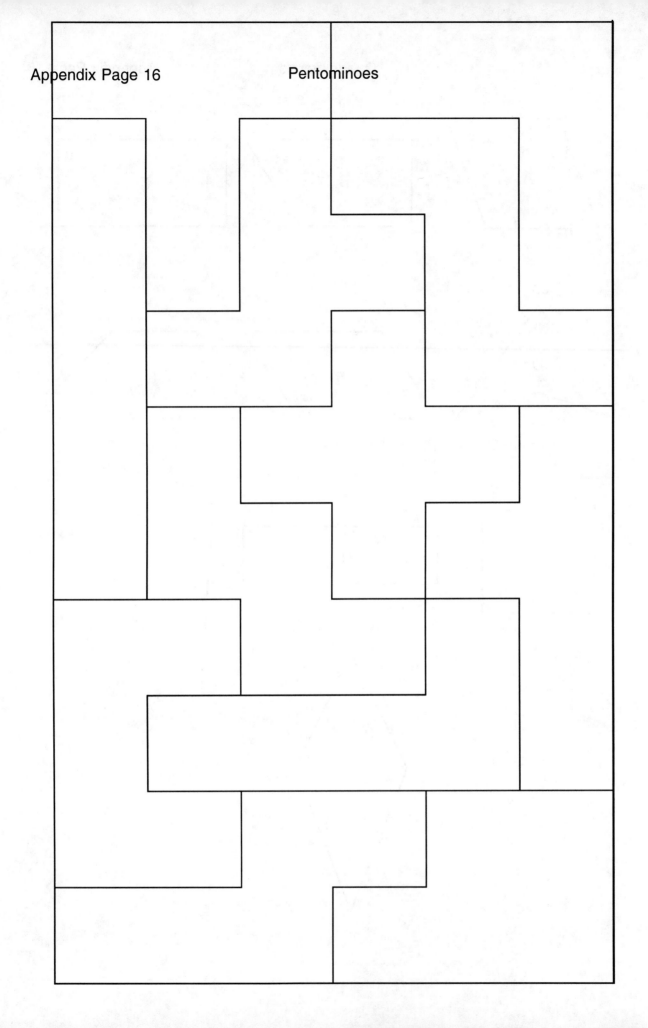

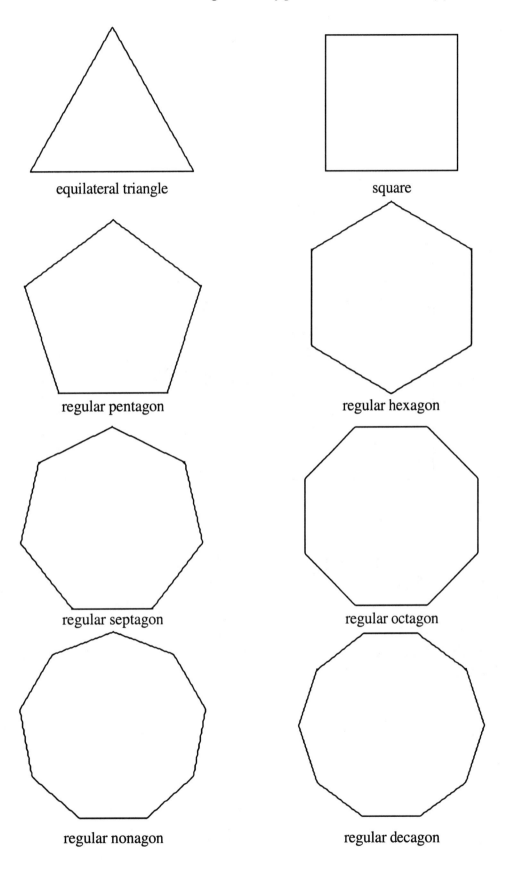

equilateral triangle

square

regular pentagon

regular hexagon

regular septagon

regular octagon

regular nonagon

regular decagon

Tangram

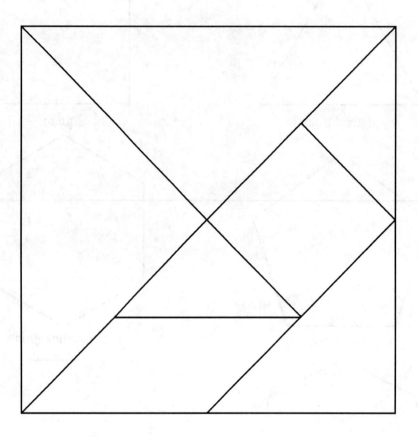

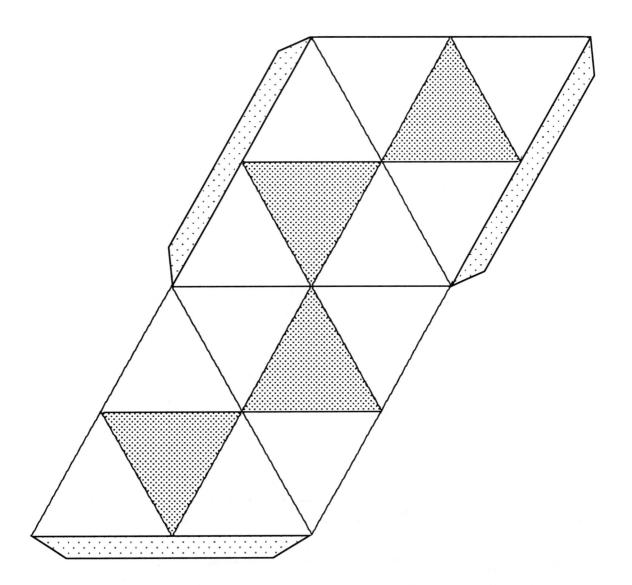

Net for Small Tetrahedra

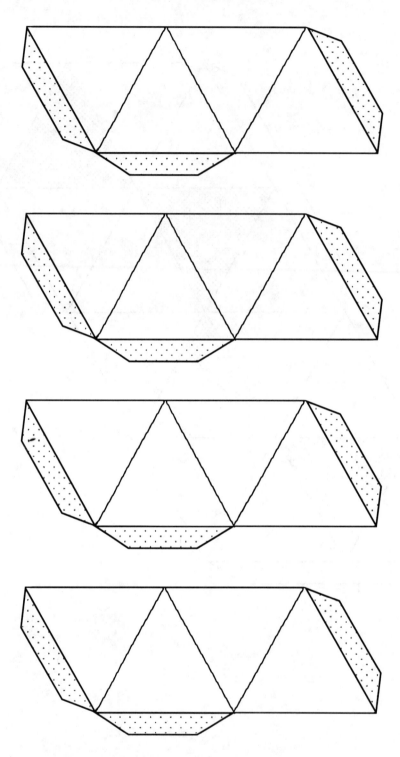

After you make the Large Tetrahedron and the Small
Tetrahedra, tape or paste the four Small Tetrahredra
to the four shaded triangles on the Large Tetrahedron
to form a Stellated Octahedron.

Net for Open-Top Prism

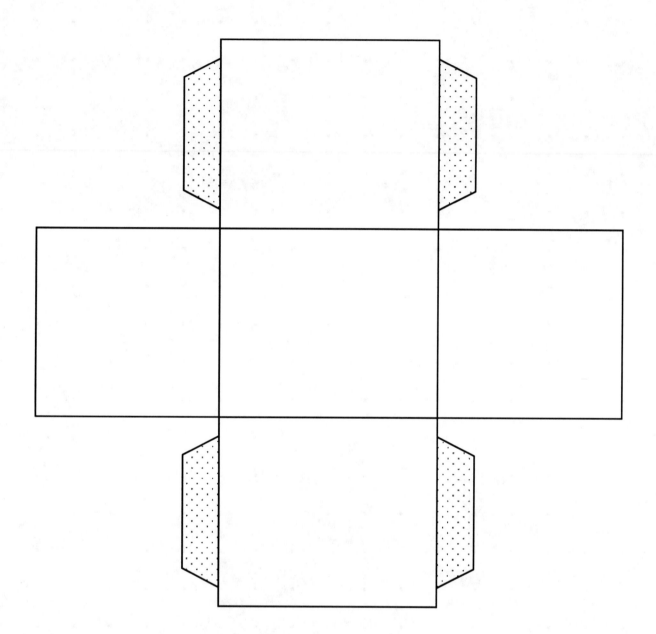

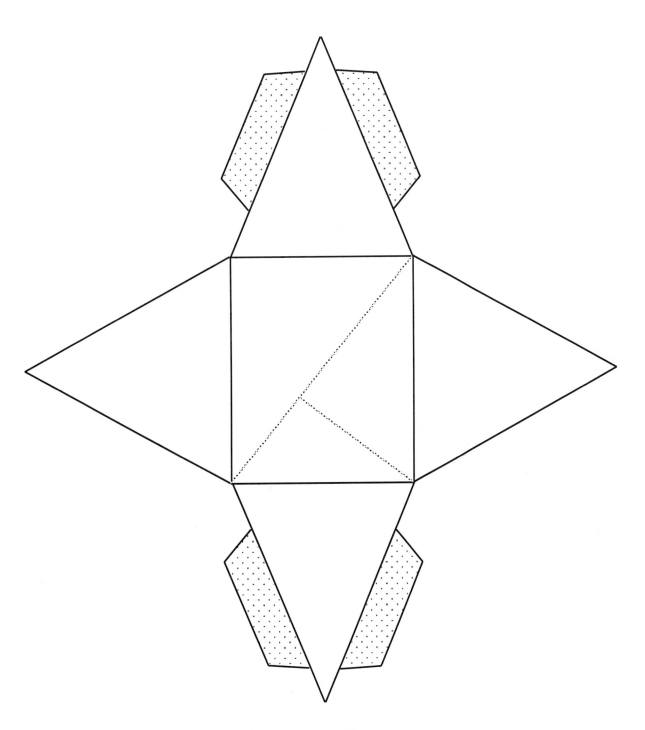

Cut along the dotted lines to open
the pyramid. Fold the two smaller
triangles inside the pyramid.

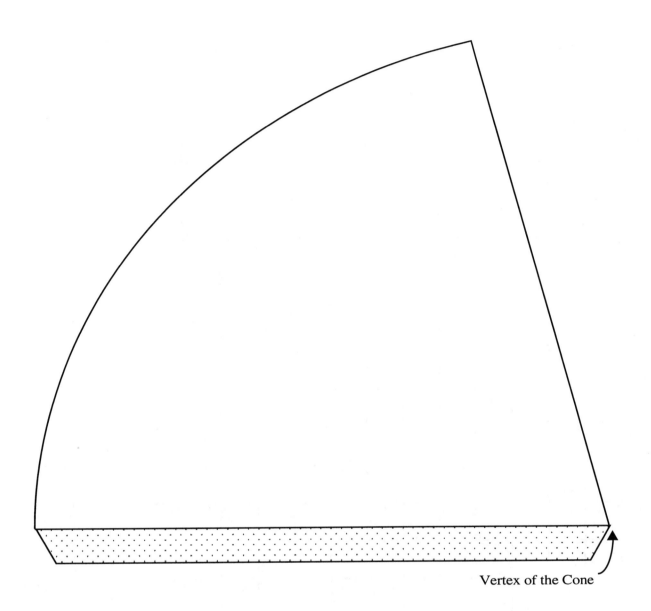

Vertex of the Cone

Centimeter Square Grid

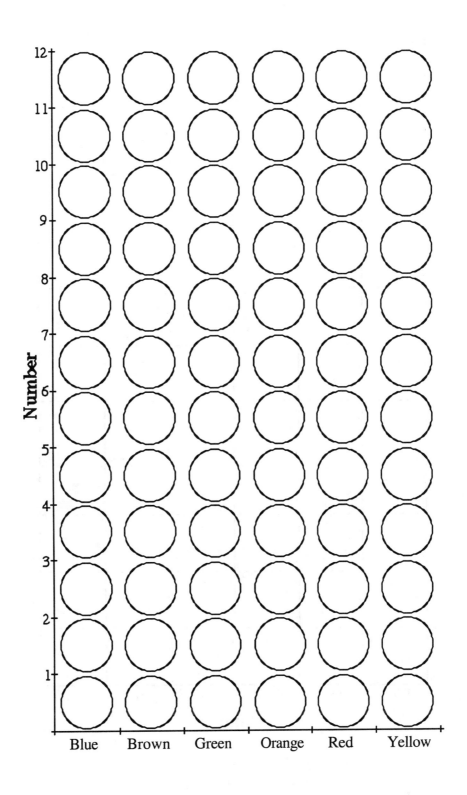

Bar Graph
M&M´s®Activity

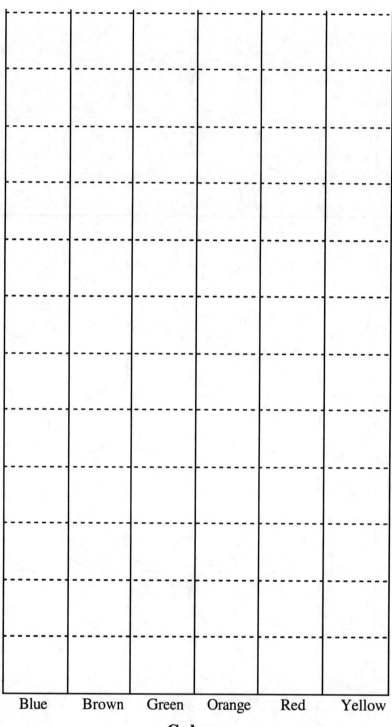

Blue   Brown   Green   Orange   Red   Yellow

Colors

Definitions:

"Geometry in the World Around Us"

1. **Angle** - the union of two rays that have the same endpoint
2. **Acute Angle** - an angle whose measure is between 0° and 90°
3. **Right Angle** - an angle made by perpendicular rays; an angle whose measure is 90°
4. **Obtuse Angle** - an angle whose measure is between 90° and 180°
5. **Zero Angle** - an angle formed by coincident rays; an angle whose measure is 0°
6. **Straight Angle** - an angle formed by opposite rays; an angle whose measure is 180°
7. **Adjacent Angles** - two angles in a plane which share a common ray and vertex but do not share any interior points; the angles must be nonstraight and nonzero
8. **Vertical Angles** - nonadjacent angles formed by two distinct intersecting lines
9. **Complementary Angles** - two angles the sum of whose measures is 90°
10. **Supplementary Angles** - two angles the sum of whose measures is 180°
11. **Parallel Lines** - two lines in a plane that either do not have any points in common or are identical
12. **Skew Lines** - two nonparallel lines in space which do not intersect
13. **Perpendicular Lines** - two lines in a plane that intersect to form right angles
14. **Simple Closed Curve** - a curve which does not touch or cross itself except where it joins at its two endpoints
15. **Polygon**
    - a simple closed curve made of three or more line segments; **or**
    - a union of three or more segments connected end to end such that each segment intersects exactly two othres at its endpoints
16. **Triangle** - a polygon with three edges
17. **Acute Triangle** - a triangle with three acute angles
18. **Right Triangle** - a triangle with one right angle
19. **Obtuse Triangle** - a triangle with one obtuse angle
20. **Isosceles Triangle** - a triangle with at least two congruent edges
21. **Equilateral Triangle** - a triangle with congruent edges
22. **Scalene Triangle** - a triangle with no edges of the same length
23. **Quadrilateral** - a polygon with four edges
24. **Parallelogram** - a quadrilateral with two pairs of parallel edges
25. **Rectangle** - a quadrilateral with right angles; a parallelogram with a right angle

Definitions (Cont'd):

"Geometry in the World Around Us"

26. **Square** - a quadrilateral with congruent edges and right angles; a rectangle with congruent edges

27. **Rhombus** - a quadrilateral with congruent edges

28. **Kite** - a quadrilateral with two distinct pairs of consecutive edges of the same length

29. **Trapezoid** - a quadrilateral with at least one pair of parallel edges

30. **Regular Polygon** - a polygon with congruent edges and congruent angles

31. **Circle** - a set of points in the plane at a given distance from a fixed point in the plane

32. **Polyhedron** - a simple closed surface in space made of four or more polygonal regions (called faces)

33. **Convex Polygon**
    - a polygon in which no interior angle has a measure greater than 180°; **or**
    - a polygon in which no line containing an edge intersects the interior of the polygon; **or**
    - a polygon in which any segment connecting points interior to or on the polygon lies completely within the polygonal region

34. **Nonconvex Polygon**
    - a polygon in which no interior angle has a measure greater than 180°; **or**
    - a polygon in which at least one line containing an edge intersects the interior of the polygon; **or**
    - a polygon in which some part of at least one segment connecting points interior to or on the polygon lies outside the polygonal region

35. **Sphere** - a set of points in space a fixed distance from a given point, called the center

36. **Cube** - a polyhedron with six congruent square faces

37. **Prism** - a polyhedron composed of two congruent polygonal faces (called bases) in parallel planes and whose lateral faces are parallelogram regions

38. **Pyramid** - a polyhedron formed by the simple polygonal region (the base), a point not in the plane of the base, and the triangular regions connecting the point and edges of the base

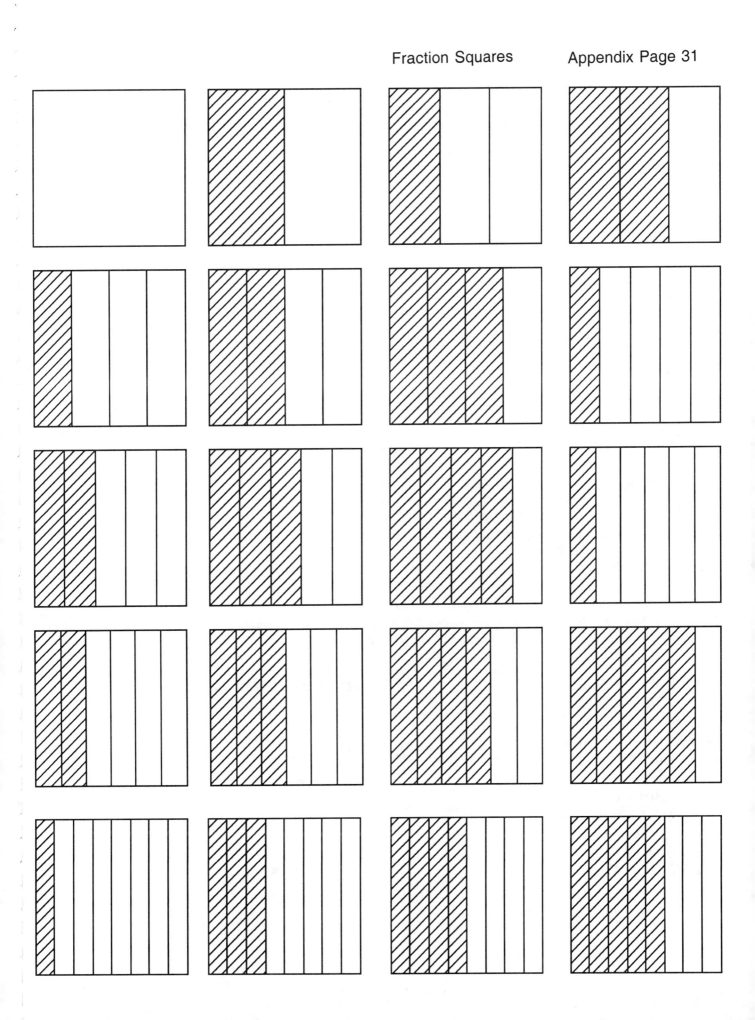

VIDEO RESOURCE LIST

Annenberg CPB
P. O. Box 2345
S. Burlington, VT 05407-2345
(800) LEARNER (532-7637)
FAX (802) 846-1850
www.learner.org
- *Against All Odds*: "What is Statistics?" and "What is Probability?"

ETA /Cuisenaire®
500 Greenview Court
Vernon Hills, IL 60061-1862
(800) 445-5985
FAX (800) 382-9326

Videotapes with Marilyn Burns:
- *Mathematics: With Manipulatives* (6 videotapes)
    Base Ten Blocks; Pattern Blocks; Cuisenaire® Rods; Color Tiles; Geoboards; Six Models
- *Mathematics: for Middle School* (3 videotapes)
    Focuses on problem solving in geometry, measurement, probability and statistics, logical reasoning, patterns and functions, algebra, and number
- *Mathematics: Teaching for Understanding* (3 videotapes)
    Models teacher-directed whole-class lessons; features small groups of students working independently on a menu of activities; shows how communication in math lessons supports children's learning and helps teachers assess students' understanding
- *Mathematics: Assessing Understanding* (3 videotapes)
    Shows how to use wait time, how to probe students' thinking and reasoning, how to follow students' leads to allow them to reveal their understanding

FASE Productions
4801 Wilshire Blvd, Suite 215
Los Angeles, CA 90010
(800) 404-3273 Ext. 367
- *Good Morning, Miss Toliver!*

Key Curriculum Press
1150 65th Street
Emeryville, CA 94608
(800) 338-7638
FAX (800) 541-2442
- *The Platonic Solids* Video
- *Stella Octangula* Video
- *Three-Dimensional Symmetry* Video

Project *MATHEMATICS!*
California Institute of Technology
Bookstore
Mail Code 1-51
Pasadena, CA 91125
(800) 514-BOOK
FAX 626-795-3156

Videotapes with Tom Apostol:
- *The Theorem of Pythagoras*
- *The Story of Pi*
- *Similarity*

PRINCIPLES AND STANDARDS FOR SCHOOL MATHEMATICS

Within NCTM's *Principles and Standards for School Mathematics,* the Principles describe particular features of high-quality mathematics education. The Standards describe the mathematical content and processes that students should learn. Together, these Principles and Standards constitute a vision to guide all of us as we work toward their implementation.

The six PRINCIPLES for school mathematics address overarching themes:

- *Equity.* Excellence in mathematics education requires equity—high expectations and strong support for all students.
- *Curriculum.* A curriculum is more than a collection of activities: it must be coherent, focused on important mathematics, and well articulated across the grades.
- *Teaching.* Effective mathematics teaching requires understanding what students know and need to learn and then challenging and supporting them to learn it well.
- *Learning.* Students must learn mathematics with understanding, actively building new knowledge from experience and prior knowledge.
- *Assessment.* Assessment should support the learning of important mathematics and furnish useful information to both teachers and students.
- *Technology.* Technology is essential in teaching and learning mathematics; it influences the mathematics that is taught and enhances students' learning.

NUMBER AND OPERATIONS STANDARD

Instructional programs from prekindergarten through grade 12 should enable all students to—

Understand numbers, ways of representing numbers, relationships among numbers, and number systems.

In prekindergarten through grade 2 all students should–
- count with understanding and recognize "how many" in sets of objects;
- use multiple models to develop initial understandings of place value and the base-ten number system;
- develop understanding of the relative position and magnitude of whole numbers and of ordinal and cardinal numbers and their connections;
- develop a sense of whole numbers and represent and use them in flexible ways, including relating, composing, and decomposing numbers;
- connect number words and numerals to the quantities they represent, using various physical models and representations;
- understand and represent commonly used fractions, such as 1/4, 1/3, and 1/2.

In grades 3–5 all students should–
- understand the place-value structure of the base-ten number system and be able to represent and compare whole numbers and decimals;
- recognize equivalent representations for the same number and generate them by decomposing and composing numbers;
- develop understanding of fractions as parts of unit wholes, as parts of a collection, as locations on number lines, and as divisions of whole numbers;
- use models, benchmarks, and equivalent forms to judge the size of fractions;
- recognize and generate equivalent forms of commonly used fractions, decimals, and percents;
- explore numbers less than 0 by extending the number line and through familiar applications;
- describe classes of numbers according to characteristics such as the nature of their factors.

In grades 6–8 all students should–
- work flexibly with fractions, decimals, and percents to solve problems;
- compare and order fractions, decimals, and percents efficiently and find their approximate locations on a number line;
- develop meaning for percents greater than 100 and less than 1;
- understand and use ratios and proportions to represent quantitative relationships;
- develop an understanding of large numbers and recognize and appropriately use exponential, scientific, and calculator notation;
- use factors, multiples, prime factorization, and relatively prime numbers to solve problems; develop meaning for integers and represent and compare quantities with them.

In grades 9–12 all students should–
- develop a deeper understanding of very large and very small numbers and of various representations of them;
- compare and contrast the properties of numbers and number systems, including the rational and real numbers, and understand complex numbers as solutions to quadratic equations that do not have real solutions;
- understand vectors and matrices as systems that have some of the properties of the real-number system;
- use number-theory arguments to justify relationships involving whole numbers.

Understand meanings of operations and how they relate to one another

In prekindergarten through grade 2 all students should–
- understand various meanings of addition and subtraction of whole numbers and the relationship between the two operations;
- understand the effects of adding and subtracting whole numbers;
- understand situations that entail multiplication and division, such as equal groupings of objects and sharing equally.

In grades 3–5 all students should–
- understand various meanings of multiplication and division;
- understand the effects of multiplying and dividing whole numbers;
- identify and use relationships between operations, such as division as the inverse of multiplication, to solve problems;
- understand and use properties of operations, such as the distributivity of multiplication over addition.

In grades 6–8 all students should–
- understand the meaning and effects of arithmetic operations with fractions, decimals, and integers;
- use the associative and commutative properties of addition and multiplication and the distributive property of multiplication over addition to simplify computations with integers, fractions, and decimals;
- understand and use the inverse relationships of addition and subtraction, multiplication and division, and squaring and finding square roots to simplify computations and solve problems.

In grades 9–12 all students should–
- judge the effects of such operations as multiplication, division, and computing powers and roots on the magnitudes of quantities;
- develop an understanding of properties of, and representations for, the addition and multiplication of vectors and matrices;
- develop an understanding of permutations and combinations as counting techniques.

Compute fluently and make reasonable estimates

In prekindergarten through grade 2 all students should–
- develop and use strategies for whole-number computations, with a focus on addition and subtraction;
- develop fluency with basic number combinations for addition and subtraction;
- use a variety of methods and tools to compute, including objects, mental computation, estimation, paper and pencil, and calculators.

In grades 3–5 all students should–
- develop fluency with basic number combinations for multiplication and division and use these combinations to mentally compute related problems, such as 30×50;
- develop fluency in adding, subtracting, multiplying, and dividing whole numbers;
- develop and use strategies to estimate the results of whole-number computations and to judge the reasonableness of such results;
- develop and use strategies to estimate computations involving fractions and decimals in situations relevant to students' experience;
- use visual models, benchmarks, and equivalent forms to add and subtract commonly used fractions and decimals;
- select appropriate methods and tools for computing with whole numbers from among mental computation, estimation, calculators, and paper and pencil according to the context and nature of the computation and use the selected method or tools.

In grades 6–8 all students should–
- select appropriate methods and tools for computing with fractions and decimals from among mental computation, estimation, calculators or computers, and paper and pencil, depending on the situation, and apply the selected methods;
- develop and analyze algorithms for computing with fractions, decimals, and integers and develop fluency in their use;
- develop and use strategies to estimate the results of rational-number computations and judge the reasonableness of the results;
- develop, analyze, and explain methods for solving problems involving proportions, such as scaling and finding equivalent ratios.

In grades 9–12 all students should–
- develop fluency in operations with real numbers, vectors, and matrices, using mental computation or paper-and-pencil calculations for simple cases and technology for more complicated cases.
- judge the reasonableness of numerical computations and their results.

ALGEBRA STANDARD

Instructional programs from prekindergarten through grade 12 should enable all students to—

Understand patterns, relations, and functions

In prekindergarten through grade 2 all students should–
- sort, classify, and order objects by size, number, and other properties;
- recognize, describe, and extend patterns such as sequences of sounds and shapes or simple numeric patterns and translate from one representation to another;
- analyze how both repeating and growing patterns are generated.

In grades 3–5 all students should–
- describe, extend, and make generalizations about geometric and numeric patterns;
- represent and analyze patterns and functions, using words, tables, and graphs.

In grades 6–8 all students should–
- represent, analyze, and generalize a variety of patterns with tables, graphs, words, and, when possible, symbolic rules;
- relate and compare different forms of representation for a relationship;
- identify functions as linear or nonlinear and contrast their properties from tables, graphs, or equations.

In grades 9–12 all students should–
- generalize patterns using explicitly defined and recursively defined functions;
- understand relations and functions and select, convert flexibly among, and use various representations for them;

- analyze functions of one variable by investigating rates of change, intercepts, zeros, asymptotes, and local and global behavior;
- understand and perform transformations such as arithmetically combining, composing, and inverting commonly used functions, using technology to perform such operations on more-complicated symbolic expressions;
- understand and compare the properties of classes of functions, including exponential, polynomial, rational, logarithmic, and periodic functions;
- interpret representations of functions of two variables

Represent and analyze mathematical situations and structures using algebraic symbols

In prekindergarten through grade 2 all students should–
- illustrate general principles and properties of operations, such as commutativity, using specific numbers;
- use concrete, pictorial, and verbal representations to develop an understanding of invented and conventional symbolic notations.

In grades 3–5 all students should–
- identify such properties as commutativity, associativity, and distributivity and use them to compute with whole numbers;
- represent the idea of a variable as an unknown quantity using a letter or a symbol;
- express mathematical relationships using equations.

In grades 6–8 all students should–
- develop an initial conceptual understanding of different uses of variables;
- explore relationships between symbolic expressions and graphs of lines, paying particular attention to the meaning of intercept and slope;
- use symbolic algebra to represent situations and to solve problems, especially those that involve linear relationships;
- recognize and generate equivalent forms for simple algebraic expressions and solve linear equations

In grades 9–12 all students should–
- understand the meaning of equivalent forms of expressions, equations, inequalities, and relations;
- write equivalent forms of equations, inequalities, and systems of equations and solve them with fluency—mentally or with paper and pencil in simple cases and using technology in all cases;
- use symbolic algebra to represent and explain mathematical relationships;
- use a variety of symbolic representations, including recursive and parametric equations, for functions and relations;
- judge the meaning, utility, and reasonableness of the results of symbol manipulations, including those carried out by technology.

Use mathematical models to represent and understand quantitative relationships

In prekindergarten through grade 2 all students should–
- model situations that involve the addition and subtraction of whole numbers, using objects, pictures, and symbols.

In grades 3–5 all students should–
- model problem situations with objects and use representations such as graphs, tables, and equations to draw conclusions.

In grades 6–8 all students should–
- model and solve contextualized problems using various representations, such as graphs, tables, and equations.

In grades 9–12 all students should–
- identify essential quantitative relationships in a situation and determine the class or classes of functions that might model the relationships;

- use symbolic expressions, including iterative and recursive forms, to represent relationships arising from various contexts;
- draw reasonable conclusions about a situation being modeled.

Analyze change in various contexts

In prekindergarten through grade 2 all students should–
- describe qualitative change, such as a student's growing taller;
- describe quantitative change, such as a student's growing two inches in one year.

In grades 3–5 all students should–
- investigate how a change in one variable relates to a change in a second variable;
- identify and describe situations with constant or varying rates of change and compare them.

In grades 6–8 all students should–
- use graphs to analyze the nature of changes in quantities in linear relationships.

In grades 9–12 all students should–
- approximate and interpret rates of change from graphical and numerical data.

GEOMETRY STANDARD

Instructional programs from prekindergarten through grade 12 should enable all students to—

Analyze characteristics and properties of two- and three-dimensional geometric shapes and develop mathematical arguments about geometric relationships

In prekindergarten through grade 2 all students should–
- recognize, name, build, draw, compare, and sort two- and three-dimensional shapes;
- describe attributes and parts of two- and three-dimensional shapes;
- investigate and predict the results of putting together and taking apart two- and three-dimensional shapes.

In grades 3–5 all students should–
- identify, compare, and analyze attributes of two- and three-dimensional shapes and develop vocabulary to describe the attributes;
- classify two- and three-dimensional shapes according to their properties and develop definitions of classes of shapes such as triangles and pyramids;
- investigate, describe, and reason about the results of subdividing, combining, and transforming shapes;
- explore congruence and similarity;
- make and test conjectures about geometric properties and relationships and develop logical arguments to justify conclusions.

In grades 6–8 all students should–
- precisely describe, classify, and understand relationships among types of two- and three-dimensional objects using their defining properties;
- understand relationships among the angles, side lengths, perimeters, areas, and volumes of similar objects;
- create and critique inductive and deductive arguments concerning geometric ideas and relationships, such as congruence, similarity, and the Pythagorean relationship.

In grades 9–12 all students should–
- analyze properties and determine attributes of two- and three-dimensional objects;
- explore relationships (including congruence and similarity) among classes of two- and three-dimensional geometric objects, make and test conjectures about them, and solve problems involving them;
- establish the validity of geometric conjectures using deduction, prove theorems, and critique arguments made by others;
- use trigonometric relationships to determine lengths and angle measures.

Specify locations and describe spatial relationships using coordinate geometry and other representational systems

In prekindergarten through grade 2 all students should–
- describe, name, and interpret relative positions in space and apply ideas about relative position;
- describe, name, and interpret direction and distance in navigating space and apply ideas about direction and distance;
- find and name locations with simple relationships such as "near to" and in coordinate systems such as maps.

In grades 3–5 all students should–
- describe location and movement using common language and geometric vocabulary;
- make and use coordinate systems to specify locations and to describe paths;
- find the distance between points along horizontal and vertical lines of a coordinate system.

In grades 6–8 all students should–
- use coordinate geometry to represent and examine the properties of geometric shapes;
- use coordinate geometry to examine special geometric shapes, such as regular polygons or those with pairs of parallel or perpendicular sides.

In grades 9–12 all students should–
- use Cartesian coordinates and other coordinate systems, such as navigational, polar, or spherical systems, to analyze geometric situations;
- investigate conjectures and solve problems involving two- and three-dimensional objects represented with Cartesian coordinates.

Apply transformations and use symmetry to analyze mathematical situations

In prekindergarten through grade 2 all students should–
- recognize and apply slides, flips, and turns;
- recognize and create shapes that have symmetry.

In grades 3–5 all students should–
- predict and describe the results of sliding, flipping, and turning two-dimensional shapes;
- describe a motion or a series of motions that will show that two shapes are congruent;
- identify and describe line and rotational symmetry in two- and three-dimensional shapes and designs.

In grades 6–8 all students should–
- describe sizes, positions, and orientations of shapes under informal transformations such as flips, turns, slides, and scaling;
- examine the congruence, similarity, and line or rotational symmetry of objects using transformations.

In grades 9–12 all students should–
- understand and represent translations, reflections, rotations, and dilations of objects in the plane by using sketches, coordinates, vectors, function notation, and matrices;
- use various representations to help understand the effects of simple transformations and their compositions.

Use visualization, spatial reasoning, and geometric modeling to solve problems

In prekindergarten through grade 2 all students should–
- create mental images of geometric shapes using spatial memory and spatial visualization;
- recognize and represent shapes from different perspectives;
- relate ideas in geometry to ideas in number and measurement;
- recognize geometric shapes and structures in the environment and specify their location.

In grades 3–5 all students should–
- build and draw geometric objects;
- create and describe mental images of objects, patterns, and paths;

- identify and build a three-dimensional object from two-dimensional representations of that object;
- identify and draw a two-dimensional representation of a three-dimensional object;
- use geometric models to solve problems in other areas of mathematics, such as number and measurement;
- recognize geometric ideas and relationships and apply them to other disciplines and to problems that arise in the classroom or in everyday life.

In grades 6–8 all students should–
- draw geometric objects with specified properties, such as side lengths or angle measures;
- use two-dimensional representations of three-dimensional objects to visualize and solve problems such as those involving surface area and volume;
- use visual tools such as networks to represent and solve problems;
- use geometric models to represent and explain numerical and algebraic relationships;
- recognize and apply geometric ideas and relationships in areas outside the mathematics classroom, such as art, science, and everyday life.

In grades 9–12 all students should–
- draw and construct representations of two- and three-dimensional geometric objects using a variety of tools;
- visualize three-dimensional objects and spaces from different perspectives and analyze their cross sections;
- use vertex-edge graphs to model and solve problems;
- use geometric models to gain insights into, and answer questions in, other areas of mathematics;
- use geometric ideas to solve problems in, and gain insights into, other disciplines and other areas of interest such as art and architecture.

MEASUREMENT STANDARD
Instructional programs from prekindergarten through grade 12 should enable all students to—

Understand measurable attributes of objects and the units, systems, and processes of measurement

In prekindergarten through grade 2 all students should–
- recognize the attributes of length, volume, weight, area, and time;
- compare and order objects according to these attributes;
- understand how to measure using nonstandard and standard units;
- select an appropriate unit and tool for the attribute being measured.

In grades 3–5 all students should–
- understand such attributes as length, area, weight, volume, and size of angle and select the appropriate type of unit for measuring each attribute;
- understand the need for measuring with standard units and become familiar with standard units in the customary and metric systems;
- carry out simple unit conversions, such as from centimeters to meters, within a system of measurement;
- understand that measurements are approximations and how differences in units affect precision;
- explore what happens to measurements of a two-dimensional shape such as its perimeter and area when the shape is changed in some way.

In grades 6–8 all students should–
- understand both metric and customary systems of measurement;
- understand relationships among units and convert from one unit to another within the same system;
- understand, select, and use units of appropriate size and type to measure angles, perimeter, area, surface area, and volume.

In grades 9–12 all students should–
- make decisions about units and scales that are appropriate for problem situations involving measurement.

Apply appropriate techniques, tools, and formulas to determine measurements.

In prekindergarten through grade 2 all students should–
- measure with multiple copies of units of the same size, such as paper clips laid end to end;
- use repetition of a single unit to measure something larger than the unit, for instance, measuring the length of a room with a single meterstick;
- use tools to measure;
- develop common referents for measures to make comparisons and estimates.

In grades 3–5 all students should–
- develop strategies for estimating the perimeters, areas, and volumes of irregular shapes;
- select and apply appropriate standard units and tools to measure length, area, volume, weight, time, temperature, and the size of angles;
- select and use benchmarks to estimate measurements;
- develop, understand, and use formulas to find the area of rectangles and related triangles and parallelograms;
- develop strategies to determine the surface areas and volumes of rectangular solids.

In grades 6–8 all students should–
- use common benchmarks to select appropriate methods for estimating measurements;
- select and apply techniques and tools to accurately find length, area, volume, and angle measures to appropriate levels of precision;
- develop and use formulas to determine the circumference of circles and the area of triangles, parallelograms, trapezoids, and circles and develop strategies to find the area of more-complex shapes;
- develop strategies to determine the surface area and volume of selected prisms, pyramids, and cylinders;
- solve problems involving scale factors, using ratio and proportion;
- solve simple problems involving rates and derived measurements for such attributes as velocity and density.

In grades 9–12 all students should–
- analyze precision, accuracy, and approximate error in measurement situations;
- understand and use formulas for the area, surface area, and volume of geometric figures, including cones, spheres, and cylinders;
- apply informal concepts of successive approximation, upper and lower bounds, and limit in measurement situations;
- use unit analysis to check measurement computations.

DATA ANALYSIS AND PROBABILITY STANDARD
Instructional programs from prekindergarten through grade 12 should enable all students to—

Formulate questions that can be addressed with data and collect, organize, and display relevant data to answer them

In prekindergarten through grade 2 all students should–
- pose questions and gather data about themselves and their surroundings;
- sort and classify objects according to their attributes and organize data about the objects;
- represent data using concrete objects, pictures, and graphs.

In grades 3–5 all students should–
- design investigations to address a question and consider how data-collection methods affect the nature of the data set;
- collect data using observations, surveys, and experiments;
- represent data using tables and graphs such as line plots, bar graphs, and line graphs;
- recognize the differences in representing categorical and numerical data.

In grades 6–8 all students should–
- formulate questions, design studies, and collect data about a characteristic shared by two populations or different characteristics within one population;

- select, create, and use appropriate graphical representations of data, including histograms, box plots, and scatterplots.

In grades 9–12 all students should–
- understand the differences among various kinds of studies and which types of inferences can legitimately be drawn from each;
- know the characteristics of well-designed studies, including the role of randomization in surveys and experiments;
- understand the meaning of measurement data and categorical data, of univariate and bivariate data, and of the term variable;
- understand histograms, parallel box plots, and scatterplots and use them to display data;
- compute basic statistics and understand the distinction between a statistic and a parameter.

Select and use appropriate statistical methods to analyze data

In prekindergarten through grade 2 all students should–
- describe parts of the data and the set of data as a whole to determine what the data show.

In grades 3–5 all students should–
- describe the shape and important features of a set of data and compare related data sets, with an emphasis on how the data are distributed;
- use measures of center, focusing on the median, and understand what each does and does not indicate about the data set;
- compare different representations of the same data and evaluate how well each representation shows important aspects of the data.

In grades 6–8 all students should–
- find, use, and interpret measures of center and spread, including mean and interquartile range;
- discuss and understand the correspondence between data sets and their graphical representations, especially histograms, stem-and-leaf plots, box plots, and scatterplots.

In grades 9–12 all students should–
- for univariate measurement data, be able to display the distribution, describe its shape, and select and calculate summary statistics;
- for bivariate measurement data, be able to display a scatterplot, describe its shape, and determine regression coefficients, regression equations, and correlation coefficients using technological tools;
- display and discuss bivariate data where at least one variable is categorical;
- recognize how linear transformations of univariate data affect shape, center, and spread;
- identify trends in bivariate data and find functions that model the data or transform the data so that they can be modeled.

Develop and evaluate inferences and predictions that are based on data

In prekindergarten through grade 2 all students should–
- discuss events related to students' experiences as likely or unlikely.

In grades 3–5 all students should–
- propose and justify conclusions and predictions that are based on data and design studies to further investigate the conclusions or predictions.

In grades 6–8 all students should–
- use observations about differences between two or more samples to make conjectures about the populations from which the samples were taken;
- make conjectures about possible relationships between two characteristics of a sample on the basis of scatterplots of the data and approximate lines of fit;
- use conjectures to formulate new questions and plan new studies to answer them.

In grades 9–12 all students should–
- use simulations to explore the variability of sample statistics from a known population and to construct sampling distributions;
- understand how sample statistics reflect the values of population parameters and use sampling distributions as the basis for informal inference;
- evaluate published reports that are based on data by examining the design of the study, the appropriateness of the data analysis, and the validity of conclusions;
- understand how basic statistical techniques are used to monitor process characteristics in the workplace.

Understand and apply basic concepts of probability

In grades 3–5 all students should–
- describe events as likely or unlikely and discuss the degree of likelihood using such words as *certain, equally likely,* and *impossible;*
- predict the probability of outcomes of simple experiments and test the predictions;
- understand that the measure of the likelihood of an event can be represented by a number from 0 to 1.

In grades 6–8 all students should–
- understand and use appropriate terminology to describe complementary and mutually exclusive events;
- use proportionality and a basic understanding of probability to make and test conjectures about the results of experiments and simulations;
- compute probabilities for simple compound events, using such methods as organized lists, tree diagrams, and area models.

In grades 9–12 all students should–
- understand the concepts of sample space and probability distribution and construct sample spaces and distributions in simple cases;
- use simulations to construct empirical probability distributions;
- compute and interpret the expected value of random variables in simple cases;
- understand the concepts of conditional probability and independent events;
- understand how to compute the probability of a compound event.

PROBLEM SOLVING STANDARD
Instructional programs from prekindergarten through grade 12 should enable all students to—
- Build new mathematical knowledge through problem solving
- Solve problems that arise in mathematics and in other contexts
- Apply and adapt a variety of appropriate strategies to solve problems
- Monitor and reflect on the process of mathematical problem solving

REASONING AND PROOF STANDARD
Instructional programs from prekindergarten through grade 12 should enable all students to—
- Recognize reasoning and proof as fundamental aspects of mathematics
- Make and investigate mathematical conjectures
- Develop and evaluate mathematical arguments and proofs
- Select and use various types of reasoning and methods of proof

COMMUNICATION STANDARD
Instructional programs from prekindergarten through grade 12 should enable all students to—
- Organize and consolidate their mathematical thinking through communication
- Communicate their mathematical thinking coherently and clearly to peers, teachers, and others
- Analyze and evaluate the mathematical thinking and strategies of others;
- Use the language of mathematics to express mathematical ideas precisely.

CONNECTIONS STANDARD

Instructional programs from prekindergarten through grade 12 should enable all students to—

- Recognize and use connections among mathematical ideas
- Understand how mathematical ideas interconnect and build on one another to produce a coherent whole
- Recognize and apply mathematics in contexts outside of mathematics

REPRESENTATION STANDARD

Instructional programs from prekindergarten through grade 12 should enable all students to—

- Create and use representations to organize, record, and communicate mathematical ideas
- Select, apply, and translate among mathematical representations to solve problems
- Use representations to model and interpret physical, social, and mathematical phenomena

Index

Index

Index

Index

Index

Index